Spillway Design – Step by Step

T0093425

Spillway Design – Step by Step

Geraldo Magela Pereira

CRC Press
Taylor & Francis Group
Boca Raton London New York

CRC Press is an imprint of the
Taylor & Francis Group, an **informa** business

A BALKEMA BOOK

Cover photo description: front cover: Tucuruí Spillway (Brazil); Monticello Spillway (Napa, USA); Bonfante Spillway (Brazil); back cover: Monjolinho Spillway (Brazil).

Published by:
CRC Press/Balkema
P.O. Box 447, 2300 AK Leiden, The Netherlands
e-mail: Pub.NL@taylorandfrancis.com
www.crcpress.com – www.taylorandfrancis.com

First issued in paperback 2021

© 2020 by Taylor & Francis Group, LLC
CRC Press/Balkema is an imprint of the Taylor & Francis Group, an informa business

No claim to original U.S. Government works

ISBN 13: 978-1-03-224025-1 (pbk)
ISBN 13: 978-0-367-41936-3 (hbk)

DOI: https://doi.org/ 10.1201/9780367816902

This book contains information obtained from authentic and highly regarded sources. Reasonable efforts have been made to publish reliable data and information, but the author and publisher cannot assume responsibility for the validity of all materials or the consequences of their use. The authors and publishers have attempted to trace the copyright holders of all material reproduced in this publication and apologize to copyright holders if permission to publish in this form has not been obtained. If any copyright material has not been acknowledged please write and let us know so we may rectify in any future reprint.

Except as permitted under U.S. Copyright Law, no part of this book may be reprinted, reproduced, transmitted, or utilized in any form by any electronic, mechanical, or other means, now known or hereafter invented, including photocopying, microfilming, and recording, or in any information storage or retrieval system, without written permission from the publishers.

For permission to photocopy or use material electronically from this work, please access www.copyright.com (http://www.copyright.com/) or contact the Copyright Clearance Center, Inc. (CCC), 222 Rosewood Drive, Danvers, MA 01923, 978-750-8400. CCC is a not-for-profit organization that provides licenses and registration for a variety of users. For organizations that have been granted a photocopy license by the CCC, a separate system of payment has been arranged.

Trademark Notice: Product or corporate names may be trademarks or registered trademarks, and are used only for identification and explanation without intent to infringe.

Publisher's Note
The publisher has gone to great lengths to ensure the quality of this reprint but points out that some imperfections in the original copies may be apparent.

Visit the Taylor & Francis Web site at
http://www.taylorandfrancis.com

and the CRC Press Web site at
http://www.crcpress.com

Typeset by Apex CoVantage, LLC

Although all care is taken to ensure integrity and the quality of this publication and the information herein, no responsibility is assumed by the publishers nor the author for any damage to the property or persons as a result of operation or use of this publication and/or the information contained herein.

Library of Congress Cataloging-in-Publication Data

This book is dedicated to my wife, Cristine, love of my life, for the constant and perennial support during my career in all my jobs; and to my daughters, Ligia and Marina, and to my granddaughter Luiza, the reason for my life.

Contents

Preface

The layout of a hydroelectric power plant (HPP) includes a main dam, a spillway, an intake, a powerhouse and the tailwater channel, and it varies from site to site depending on the topographical and geological characteristics of each site, which strongly condition the project.

This book presents the methodology of *Spillway Design – Step by Step* in a simple, objective and practical way, with more details than that were presented in *Hydropower Plants Design – Step by Step* (2015), by the same author. It should be noted that the terms of the Technical Dictionary on Dams from the International Commission on Large Dams (ICOLD) were observed in this book.

The spillway is the structure to outflow the flows that exceed those that will be turbinated and which, therefore, cannot be stored in the reservoir. In other words, the spillway is the structure used to prevent the absolute maximum water level of the reservoir from being exceeded, to allow the freeboard to be preserved, in order to protect the dam against overtopping. The discharged flows are returned to the riverbed, routinely, by an artificial channel.

According to USBR (1983) and ICOLD (1995), the main causes of dam breaks are: foundation failure, 40%; inadequate spillway, 23%; poor construction, 12%; uneven settlement, 10%; high pore pressure, 5%; acts of war, 3%; embankment slips, 2%; defective materials, 2%; incorrect operation, 2%; and earthquakes 1%. This statistic shows that: (1) the foundation must be well investigated; (2) the design flood must be well evaluated; (3) the construction must be well done by a specialized company with an unblemished reputation, with adequate materials; and (4) the plant must be operated according to the preestablished rules.

So the importance of a safe spillway cannot be underestimated. The percentage relative to an inadequate spillway includes those resulting from the inadequate operation of the gates for some reason (Leyland 2008, see section 9.2.8). Extensive discharge capacity is of paramount importance for the safety of earth and rockfill dams, which are likely to be destroyed if they are overtopped.

The selection of the design flood must be made accordingly ICOLD (1987). In Brazil this flood should be estimated according to Eletrobras/CBDB (2003). For dams with heights greater than 30 m, or whose collapse involves the risk of loss of life, the design flood will be the maximum probable flood – CMP. For dams smaller than 30 m, or with a reservoir with a volume of less than 50×10^6 m^3 and with no risk of loss of life, the project flood will be defined through risk analysis, with a minimum recurrence of 1.000 years. These studies are not part of the scope of this book.

The hydraulic design of the structure has been made according to the criteria of the USBR (1948), contained in HDC (1959), DSD (1974) and ICOLD (1987). The hydraulic design of the stilling basin structure is made according to Elevatorski (1959) and HDSBED (1983).

Projects should be supported by hydraulic studies on reduced physical models. Usually two models are used:

- a three-dimensional model to analyze, among other topics: (a) if the spillway will receive the affluent flow well, with flow lines without discontinuities, detachments or excessive perturbations of the liquid vein that can cause reduction of flow capacity and/ or cavitation in the structure; (b) interferences in the contiguous structure of the water intake; (c) the safe return of the flow to the riverbed, minimizing the possibilities of scours; (d) the river diversion;
- a two-dimensional model for detailed testing of the spillway and stilling basin, as well as the hydrodynamic pressures.

It should be noted that currently is more and more in practice in Europe to use computational fluid dynamics (CFD) modeling of the spillways, as these models could be easily adjusted and modified if necessary.

Among the several examples, the first one presented in the book is that of the spillway of the Tucuruí HPP (Figure 0.1). It is one of the largest spillways in the world, designed for 110.000 m³/s. The structure is 570 m wide, 92 m high, with 23 gates 20 m wide by 21 m high. The useful width at the boarding lip of the jet is 552 m. The specific discharge is of the order of 200 m³/s/m. The flow velocity in the jet takeoff varies between 30–35 m/s. The plant started operating in October 1984. Since then, the spillway has operated frequently without any problem, as reported by Magela *et al.* (2015) and will be treated in this book.

Figure 0.1 UHE Tucuruí, Tocantins river, State of Pará, Amazon region, Brazil.

The second example is the spillway of the Cethana HPP (100 MW), Tasmania (Figure 0.2). The rockfill dam with concrete face, 213 m wide and 110 m high, was located in a narrow valley on the river Forth. The ski jump chute spillway in the left abutment and without control has a discharge capacity of 1.980 m³/s. It was completed in 1971 (Mitchell, 1973).

The book was structured in a similar way to that presented in the DSD-USBR (1960/1973/ 1974). It encompasses in some detail the most commonly used ogee spillways. Side spillways projects, morning glory spillways, labyrinth spillways and bottom spillways are presented in a simplified way. The definition and characterization of the routinely used types of spillways is discussed, presenting the criteria of choice, the definition of the arrangement (location in plan), studies of alternatives, the hydraulic design, including the terminal structure of the energy dissipator, and studies on reduced models to optimize the design (Figures 0.3 and 0.4).

Figure 0.2 Cethana HPP, Tasmania. Ski jump spillway in the left abutment.

Figure 0.3 Barra Grande HPP, Pelotas river.Brazil, State of Minas Gerais.

Figure 0.4 Irapé HPP, Jequitinhonha riverBrazil, States of Rio Grande do Sul/Santa Catarina.

The book also includes the hydrodynamic efforts on the chute, the aspects related to the cavitation in the structure, a summary on the aeration, as well as the evaluation of the possibilities of occurrence of scours in the rocky massif downstream, as well as deterioration of the structure itself. In addition, the book contemplates the mechanical equipment, gates and valves and some notes on the maintenance of the gates, the risks to the dam in cases of failure to open them for any reason, or for failure of the handling equipment during the occurrence of flooding, which can lead to overtopping and rupture of the dam (Figure 0.5).

The main aspects related to the plan of operation of the gates and of the monitoring of the performance of the structure and the equipment throughout its useful life are presented.

It also presents the case of rupture in July 17, 1995, of one radial gate of the spillway of the Folsom dam (198,7 MW) of the USBR in the state of California (USA). In the second photo of the arrangement of this plant, in Figure 0.6 we can observe the construction of the additional spillway in the left abutment near the road. The need for additional spillway in a hydroelectric because of the increase in the design flood over the useful life of the plant will be highlighted.

To illustrate the points of the spillway projects, examples from the bibliography will be presented, especially from the publication "Great Brazilian Spillways – An Overview of

Figure 0.5 Karakaya HPP, Euphrates river. Turkey.

Brazilian Practice and Experience in Projects and Construction of Spillways for Large Dams," CBDB (2002).

Large dams store large volumes of energy that, if suddenly released, make a disaster inevitable. In 1952, Andre Coyne, designer of the Malpasset dam, Fréjus, France, said that of all man-made works dams are the deadliest. Seven years later, in 1959, the Malpasset dam broke and caused extensive damage downstream, as well as 423 fatalities.

Figure 0.6 Folsom dam (California, USA).

Acknowledgments

The author wishes to thank especially the Brazilian Committee on Large Dams, in the person of its President Carlos Henrique Medeiros, for the permission to use the data of its publications.

The author thanks the following colleagues who provided information and assistance during the preparation of the book:

Civil Engineer Murilo Lustosa Lopes, Brasília, Federal District, Brazil;
Civil Engineer Sérgio Corrêa Pimenta, Florianópolis, Santa Catarina, Brazil;
Designer Keren Feitosa, Florianópolis, Santa Catarina, Brazil;
Designer Reynaldo Medeiros Brandão, Petrópolis, Rio de Janeiro, Brazil.

The author also thanks Angela Medina, Senior Affairs Specialist, Bureau of Reclamation, who responded on March 12, 2019, to my request stating that USBR publications are not copyrighted.

The author expresses his gratitude to the following people who kindly responded to his request and gave permission to use his data:

Bryan Leyland, Leyland Consultants, Auckland, New Zealand;
Hubert Chanson, PhD DEng, Professor, School of Engineering, The University of Queensland, Brisbane, Australia;
Erik Bollaert, Director of Aquavision Engineering Ltd, Lausanne, Switzerland;
Sébastian Erpicum, Head of the Laboratory of Engineering Hydraulics, University of Liège, Belgium;
Willi H. Hager, Professor Dr., VAW, D-BAUG, ETH, Zurich, Switzerland;
Antje Bornschein, Dr.-Ing., Technical University of Dresden, Faculty of Civil Engineering, Institute of Hydraulic Engineering and Technical Hydromechanics, Germany;
Cesare De Simone, Manager of Hydroplus, Rio de Janeiro, Brazil.

About the author

Geraldo Mageia Pereira, Civil Engineer, graduated from in Brasília University (July, 1974), with 40 years of experience in hydroelectric power plant projects (HPPs), 30 years of them at Engevix Engineering. Master Degree, MSc, in Civil Defense Protection at Fluminense Federal University, Niterói – Rio de Janeiro (2017). Worked in the geotechnical and hydraulic areas, including studies on reduced models, arrangements and planning and monitoring of the construction, coordination and direction of projects, in its various phases: inventory studies, feasibility studies, basic projects and executive projects. He worked in the commercial area between 1998 and 2012, developing business for the implementation of projects in EPC contracts. Published one paper in the XVI ICOLD of San Francisco (USA) in 1988, about the "Historic Flood During the 2nd Phase of Tocantins River Diversion for the Construction of Tucuruí Power Plant," Q. 63, R.2, and several papers in Brazilian seminars of large dams. Published three books: *Hydroelectric Power Plants Design Step by Step* in 2015, *Spillways Design Step by Step* in 2016 and *Accidents and Ruptures of Dams* (pdf) in 2018.

Main companies/projects

Federal Railway Network (1974/1975): 3rd line in the eastern region of the city of São Paulo between Manoel Feio and Roosevelt stations;

Hidroesb, Rio de Janeiro (1975/1976): Hydraulic Models;

Enge-Rio, Rio de Janeiro (1977/1979): Balbina HPP (250 MW), Eletronorte; MILDER-KAISER Engineering (80/81): Rosana HPP (320 MW), Cesp;

Engevix, Rio de Janeiro (1981–2012): Tucuruí HPP (1st phase – 4.000 MW), Eletronorte, including two years of training of the team of operation of the 23 gates of the spillway; Santa Isabel HPP (2.200 MW), Eletronorte; Salto da Divisa HPP and Itapebi HPP, Furnas; Canoas I HPP (72 MW) and Canoas II HPP (82,5 MW), Cesp. EPC: Capim Branco I HPP and Capim Branco II HPP – EPC and Baguari HPP – EPC.

Magela Engineering, Rio de Janeiro since October 1994:

- Design of several small hydroelectrics (PCHs) for several clients, totalizing 1,000 MW;
- Inventory Studies of the rivers Sucuriú, Verde, Iguatemi and Paraíso (MS), Araguaia/ Tocantins – revision (TO), Teles Pires (MT); Fetal and Prata (northwest of MG).
- Consultant of Engevix Engineering: the implementation of the Monte Serrat (25 MW), Bonfante (19 MW) and Santa Rosa (30 MW) SHPs in the State of Rio de Janeiro.
- Consultant of Leme Engineering, Belo Horizonte (1997) for projects in Chile (Laja I and Laja II) and in Panama (Teribe and Changuinola);
- Coordinated for Coppetec/Eletrobras the revision of the Manual of Small Hydroelectric Power Plants (PCHs-1997/1998);
- Consultant of SGH-ANEEL for analysis of SHP projects (2000);
- Consultant of CPFL Energia to develop business with SHPs (2008–2010).

Acronyms

ABGE	Brazilian Association of Engineering Geology (1968)
ABRH	Brazilian Association of Hydric Resources
ANA	National Water Agency – Brazil
ANEEL	National Electric Energy Agency – Brazil
ASCE	American Society of Civil Engineers (1852)
ASME	American Society of Mechanical Engineers (1880)
BCOLD (CBDB)	Brazilian Committee on Large Dams (1957)
CCR	Concrete Roller Compacted
CEHPAR	Center for Hydrology Professor Parigot de Souza – Brazil (1959)
CEMIG	Energy Company of Minas Gerais State – Brazil (1952)
CESP	Energy Company of São Paulo State – Brazil (1966)
CFE	Federal Electricity Commission – Mexico
CHESF	São Francisco Hydroelectric Company – Brazil (1945)
CIRIA	Construction Industry Research and Information Association (1960)
CJCE	Canadian Journal of Civil Engineering
CoE	Corps of Engineers, Department of The Army (1802)
CPFL	Paulista Company of Power and Light – Brazil (1912)
CSU	Colorado State University (USA)
DNOCS	National Department of Drought Works – Brazil
DSD	Design of Small Dams (1st ed. 1958); "Low Dams"-1938
ELETROBRAS	Brazilian Electric Power Plants – Holding (1961)
ELETRONORTE	Power Plants of Northern Brazil (1973)
EPFL	École Polytechnique de L'Université de Lausanne (1853)
FURNAS	Furnas Power Plants, Brazil (1957)
HDC	Hydraulic Design Criteria (1a edition 1952)
HDS	Hydraulic Design of Spillways (1965)
HDSBED	Hydraulic Design of Stilling Basin and Energy Dissipators (1958)
HIDROESB	Hidrotechnical Laboratory Saturnino de Brito (1966)
HPP	Hydroelectric Power Plant
IAHR	International Association of Hydraulics Research (1935)
ICE	Institute of Civil Engineers (London, 1818)
ICOLD	International Commission on Large Dams (1928)

IPH	Institute of Hydraulic Research, UFRGS, Brazil (1953)
JASA	The Journal of the Acoustical Society of America
JCD	Journal of the Construction Division (ASCE)
JHD	Journal of the Hydraulic Division (ASCE)
JHE	Journal of the Hydraulic Engineering (ASCE)
JPD	Journal of the Power Division (ASCE)
LACTEC	Institute of Technology for Development, UFPR, Brazil
LAHE.T	Laboratory of Experimental Hydraulics and Water Resources, FURNAS
LBS	Large Brazilian Spillways
LIGHT	Light Electricity Services (1905)
LNEC	National Laboratory of Civil Engineering. Lisbon (1946)
MDB	Main Brazilian Dams
NUST	Norwegian University of Science and Technology
NZSOLD	New Zealand Society of Large Dams
PMF	Probable Maximum Flow
PMP	Probable Maximum Precipitation
POG	Plan of Operation of the Gates
SHP	Small Hydroelectric Power Plant
SOGREAH	Societé Grenobloise d'Etudes et d'Applications Hydrauliques
SSARR	Streamflow Synthesis and Reservoir Regulation
TR	Recurrence time
TVA	Tennessee Valley Authority
UFPR	Federal University of Paraná, Brazil (1912)
UFRGS	Federal University of Rio Grande do Sul, Brazil (1895)
USA	United States of America
USBR	United States Bureau of Reclamation (founded in 1902 by Pres. Roosevelt)
USBR-HLT	United States Bureau of Reclamation, Hydraulic Laboratory Techniques
USCE	United States Corps of Engineers
USSD	United States Society on Dams
USGS	United States Geological Survey
USP	University of São Paulo, Brazil (1934)
WES	Waterways Experiment Station, US Army Engineer (founded in 1930)
WL	Water level
WL max	Absolute maximum water level
WL normal	Normal water level
WL jus	Downstream water level
WL res	Reservoir water level
WP&DC	Water Power & Dam Construction

Symbols

A	area (m^2);
B	width of the spillway corresponding to each aerator (m);
B	bucket width at jet takeoff – ski jump (m);
$2Bu$	width of the jet in the impact zone, Figure 6.2 (m);
C, C_d, C_o	discharge coefficient;
C_f	coefficient of friction, calculated at each point according to Equation 4.4;
C	air concentration (%);
d	depth of water on the bucket (m);
d_1	depth of water in the entrance of the basin – section 1 (m);
d_2	depth of water in the exit of the basin – section 2 (m);
d_L	depth of water at jet takeoff (m);
D	depth of scour (m), Equation 7.1;
D_{max}	maximum depth of scour (m) – Equation 5.3;
D	valve diameter (m);
dm	mean particle diameter (m);
D	effective duct area per unit of chute width – Figure 8.38: $D = CA/B$ (m^2/m);
E_1	energy Section 1 – Figure 5.39;
E_2	energy Section 2 – Figure 5.39;
E_u	energy, Equation 6.3;
F_1	Froude number at basin entrance – Equation 5.9;
g	acceleration of gravity (m^2/s);
Go	net gate opening (m) – Figure 9.14;
h	depth of water downstream (m);
h_p	pressure head against the boundary – Figure 4.4 (m);
H, H_d, H_0, H_t	hydraulic head, design head, or total head, equal to the difference between the upstream water level and the downstream water level (m);
H_1	difference between the jet takeoff elevation and the elevation of downstream water level – Figures 6.4 and 6.8 (m);
J	quantity of movement – Equations 5.1 and 6.1 (t/m);
K_a	coefficient of contraction of the walls of the spillway – Figures 3.7 and 3.8;
K_p	coefficient of contraction of the piers of the spillway – Figure 3.9;
K	concrete roughness (0,0006 m);

K	constant from Equation 7.1 – Figures 7.1 and 7.2;
K, x, y and z	constants for each of the formulas in Table 7.2;
L	hydraulic jump length; effective crest length (m);
L'	net crest length (m);
N	power spilled – Equation 5.2 (MW/m);
P	crest height of spillway sill (m);
$\sqrt{p_w^2}$	standard deviation of pressure fluctuations, calculated by Equation 4.6 (m);
P, P_m	mean pressure, Equation 4.7 (m);
Pu	pressure in the jet's impact zone (m) – Equations 6.7, 6.8 and 6.9;
p	absolute pressure at a point outside the cavitation zone, Equation 8.1 (m);
p_v	vapor pressure inside the cavitation bubbles, which is usually taken equal to the water vapor pressure at the temperature at which it is present, Equation 8.1;
Δp	the difference between the atmospheric pressure and the average pressure under the jet, as measured along the vertical face of the step or ramp, Equation 8.2 (N/m^2);
q	unit discharge (m^3/s/m);
Q	total discharge (m^3/s) – Equation 6.2;
Q_1	discharge section 1 – upstream, (m^3/s);
Q_2	discharge section 2 – downstream, (m^3/s);
Q_a	total air discharge (m^3/s);
q_a	specific air discharge per unit of chute width (m^3/s/m);
R	radius of the bucket (m);
Ru	radius of the jet in the impact zone, Figure 6.2 (m);
S	area (m^2);
t	aerator ramp height (m);
V_E	flow velocity (m/s):
V_o, V_1	approach velocity (m/s);
V_1, V_2	velocity section 1 (upstream); velocity section 2 (downstream) (m/s);
V_L, V_o	takeoff velocity of the jet (m/s);
Vu	velocity in the jet impact zone (m/s) – Figure 6.4 and Equation 6.7;
v	velocity in the boundary layer at a point distant "y" from the surface (m/s);
v_o	flow velocity (m/s);
x, y, z	exponents of Equation 7.1 – see Equation 7.2;
Yk	submerged jet core length (m) – Equations 6.5 and 6.6;
W	basin width (m);
$WLres$	reservoir water level (m);
$Wjus$	water level downstream (m);
α	angle of rotation from beginning of the curve in the bucket (^{o}C) – Figure 4.4;
$\propto_T$	total deflection angle (°C) – Figure 4.4;
α	spillway chute angle (°C) – Figure 8.38;
a_1	kinetic energy correction factor; approx. = 1 for practical problems – Figure 1.7;
$\propto_i$ or $\propto_u$	the internal angle of diffusion of the jet;

α	spillway chute angle (°C) – Figure 8.38;
β	coefficient that normally varies between 3 and 5 – Equation 4.6;
β	coefficient that measures the performance of the aerator, relation between the air and water – see Equations 8.3 to 8.7;
β	angle between the tangent to the face of the gate and the spillway sill – Figure 9.9;
θ	bucket lip angle (°C) – Figure 5.3;
X	throw distance of the jet (m), as HDC 112–8 – Figure 5.3;
γ	specific weight of water = 1,0 t/m^3;
δ	thickness of boundary layer (HDC 111–18 a 111–18/5) (m);
μ	coefficient of dynamic viscosity (kgm^{-2}s)
ϑ	coefficient of kinematic viscosity, equal to μ/ρ, (m^{2}s^{-1});
ρ	water density (t.s^{2}m^{-4});
ρ_a	air density, Equation 8.2 (kg/m^3).
σ_{CR}, σ_I	incipient cavitation index, Equation 8.1;
τ_o	shear stress in concrete (t/m^2);
ϕ	angle between the tangent to the radius and the horizontal at each point on the bucket – Figure 4.20 (m);
ϑ	aerator ramp angle – Figure 8.38.

Chapter 1

Introduction

This book presents the methodology of design of spillways of hydroelectric power plants, step by step. It is understood that the basic concepts of hydraulics engineering and of rock erosion are well known to users. These principles are set out in basic texts cited in the references, as well as in articles and books available on the internet. However, where necessary, summaries of specific points will be presented.

The projects of the structures of the hydroelectric plants, including the spillways, have been carried out in accordance with the guidelines and criteria of the US Army Corps of Engineers and USBR publications: Hydraulic Design Criteria (1959); Design of Small Dams (1974); and USACE, Hydraulic Design of Spillways (1965). Mention should be made of ICOLD Bulletin No 58, Spillways for Dams (1987). In Brazil, these criteria can be found in Eletrobras/CBDB (2003), available on the internet.

1.1 Initial considerations

The following illustrations show the Tucuruí HPP in the Tocantins river, state of Pará, Brazil. Figure 1.1A shows part of the right dam, the spillway and powerhouse and the left abutment. Figure 1.1B shows part of the spillway, the right dam and the right abutment. The dam is 7 km long, including the spillway, with a width of 560 m.

Figure 1.1A Tucuruí HPP. Approach flow to the spillway. Observe the rotation of the flow around the spigot over the dam.

Figure 1.1B Tucuruí HPP. Reservoir filling. Part of the spillway and part of the right dam. Spigot detail.

To optimize the performance of the spillway operation during its design life, say at least 50 years, one must always choose:

- **a site with good foundations conditions**, preferably with sound rock, to withstand the hydrodynamic pressures that will be applied to the riverbed by the flow coming from the spillway; and
- a layout whose works adequately receives the approach flow, with flow lines without separations, discontinuities or detachments and with a well-distributed velocity profile to reduce the pressures that will be applied to the river downstream by the effluent flow of the spillway.

The spigot on the Tucuruí right dam slope was designed exactly in order to improve the approach flow on the right side of the spillway in contact with the earth dam. The solution to the left side – a wall between the spillway and the intake – will be presented in Chapter 10.

In order to verify these aspects, two hydraulic models were used, as presented in Chapter 10 (three-dimensional and two-dimensional models, as shown in Figures 1.2 and 1.3).

Figure 1.4 shows a typical flow net; Figures 1.5 and 1.6 show details of flow separation zones.

Figure 1.2 Tucuruí HPP (Brazil). Partial view of the three-dimensional model (scale 1:150).

Figure 1.3 Tucuruí spillway. Two-dimensional model (scale 1:50). It is worth noting the good performance of the flownet over the spillway (Q = 100.000 m³/s).

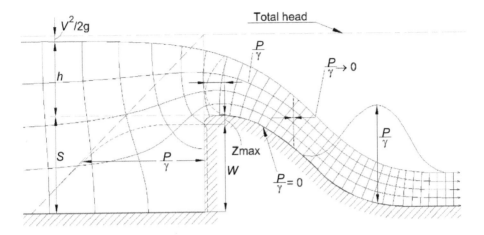

$V^2/2g$

Total head

h

$\frac{P}{\gamma}$

$\frac{P}{\gamma} \rightarrow 0$

S

$\frac{P}{\gamma}$

Zmax

W

$\frac{P}{\gamma} = 0$

$\frac{P}{\gamma}$

Figure 1.4 Typical flownet.

Source: (Rouse, 1938, 1950).

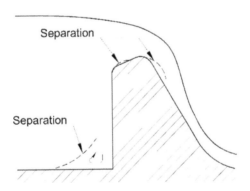

Separation

Separation

Figure 1.5 Separation zones of the flow – sketch.

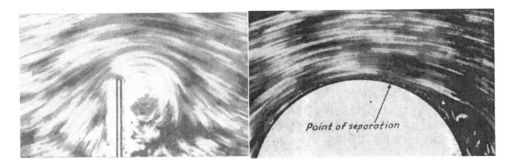

Point of separation

Figure 1.6 Pattern of flow on a plate and on a cylinder.

Source: Images (Rouse, 1938, 1950).

According to Brater and King (1976), a spillway is a notch through which water flows; it may be a depression on the side of a tank or along the contour of a reservoir or channel, or may be a spillway dam or other similar structure.

Classified in accordance with the shape of the notch, there are rectangular, triangular (or V-notch), trapezoidal and parabolic spillways (Figure 1.7).

The edge or surface over which the water flows is called the crest of the spillway. The overflowing sheet of water is termed the nappe. The depth of water producing the discharge, H, is the head.

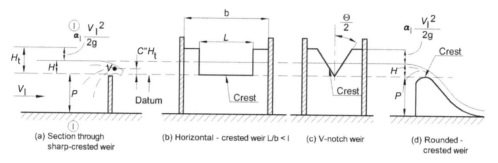

Figure 1.7 Spillways definition sketch.

Source: (Brater and King, 1976).

The main parameters are:

H_t total head (m);
H hydraulic head (m) $\Rightarrow Q = c\,L\,H^{1,5}$ (m^3/s), as will be seen in Chapter 3;
V_1 velocity in the approach channel (m/s);
V velocity over the crest (m/s);
P crest height (m), parameter that strongly influences the discharge coefficient (c);
α_1 kinetic energy correction factor $\approx 1,0$.

Sharp-crested weirs are useful only as a means of measuring flowing water. Spillways not sharp-crested are commonly incorporated in hydraulic structures.

This book covers in more detail the rounded crested weirs or ogee spillways used in hydroelectric plants.

The projects of the other types of spillways, for example, morning glory, labyrinth, siphon, stepped, etc., will be presented in a simplified way.

1.2   Phases of studies of the basin. The cases of Banqiao (China), Mascarenhas de Moraes and Itá (Brazil)

The first phase of studies of a river basin is the Hydroelectric Inventory Studies, in which are defined how many plants can be implanted in the basin, with their respective operational levels. In these studies, the layouts of future power plants are characterized according to the topographical and geological characteristics of each site.

In the next phase of studies, the Feasibility Studies, all the basic studies are deepened and better investigated: hydrological studies, energetic studies, geological and geotechnical studies, studies of alternatives of layouts, environmental studies and the electromechanical studies.

Several alternatives of layouts are usually researched, optimized and compared, aiming at the selection of only a few to be detailed in the final stage of economic-energy feasibility. The spillway type is always linked to the type of layout defined.

The floods must be estimated according to ICOLD Bulletin 142, Safe Passage of Extreme Floods (2012). In Brazil, the discharges series are defined by ANA – Agência Nacional de Águas. The maximum design flood (CMP) should be determined as recommended in "Guide to Calculate the Design Flood of Spillways Project," Eletrobras (1987) and in "Criteria of Civil Project of Hydroelectric Power Plants," Eletrobras/CBDB (2003).

The majority of dam accidents – 23% – are due to the insufficient capacity discharge of the spillways, according to ICOLD Bulletin 99 (1995). This means that the CMP may have been undersized, which occurs with some frequency for several reasons that are not the subject of this book. It is recommended that the reader consult the referenced manuals and also "Dams and Public Safety," Jansen (1983), which contains several examples of dam breaks.

Of particular note is the Banqiao dam (Figure 1.8) on the Ru river, the largest tributary of the Huai river, in Zhumandian City, Henan Province, China. More than 100 dams were built in this basin after the 1950 flood to control floods and generate electricity (Dai Qing, 1989).

This homogeneous earth dam, with clayey backrests and a sandy shale nucleus, 24,5 m high, was built from April 1951 to June 1952, with the support of Soviet consultants. Data extracted from the internet record that the dam was made of erodible material and that the compaction was deficient, which puts in doubt the geotechnical design. The capacity of the spillway was 1.742 m^3/s, for TR = 1.000 years.

According to the articles consulted, the hydrological studies of the project, due to lack of data, were developed in a lower standard than usual. The number of sliding gates (sluice gates) had been reduced from 12 to five.

Soon after the construction was completed in 1952, cracks began to appear on the dam and gates, which were soon repaired under the supervision of Soviet civil engineers. After the flood of 1954, all the reservoirs of the basin were elevated. Banqiao reservoir was raised 3 m. In August 1975, some extreme rain events occurred following the collision of typhoon Nina with a cold front. On August 8th, the critical event happened. It rained the annual total in 24 hours. The water level began to rise, reached the elevation 107,9 m.

The opening of the gates was hampered by the accumulation of sediments. The water pressure collapsed the gates instantly. At the same time, the dam was broken; it broke when the flood hit 13.000 m^3/s. Another 62 dams in the basin ruptured.

It is estimated that this rupture caused the death of 26.000 people. Another 160.000 people died of epidemics and famine. Eleven million people lost their homes. These numbers are unthinkable and unacceptable.

The dam was rebuilt 18 years later, in 1993, with a height of 50,50 m. The flood design of the spillway was redefined as 15.000 m^3/s, ten times larger than the original dam fill.

For this reason, reviewing hydrological studies of old power plants projects has become a current practice in Brazil.

It is also imperative to emphasize the importance of hydraulic studies in reduced models for the optimization of the structures of the projects.

Figure 1.8 Banqiao: rupture and reconstructed dam (China).

In the case of the Mascarenhas de Moraes HPP in the Grande river (Minas Gerais State), presented later, forty years later FURNAS decided to revise flood studies using a new methodology and concluded that it was necessary to construct an additional spillway to accommodate 3.300 m³/s more.

The following figures show the Itá HPP in the south of Brazil. This power plant was implanted in a section of the Uruguay river that makes a turn in a 180° "U.". It should be noted that this project, due to the characteristics of the site and for economic reasons, has two spillways.

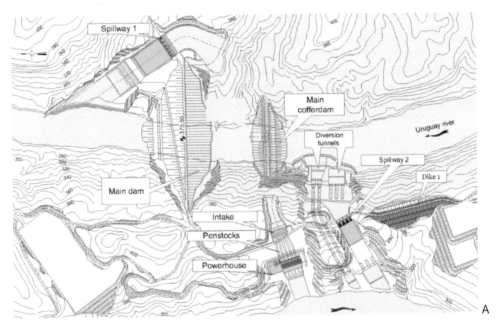

Figure 1.9 Itá HPP. Santa Catarina/Rio Grande do Sul, Brazil. CBDB/MBD (2009).A – Layout of the power plant. B – Aerial view of the works.

Beyond the definition of the design flood, it is important to program enough surveys to investigate the area of the dam and of the spillway. It's necessary to characterize in detail the rock foundation in the region of the energy dissipation structures to subsidize the Feasibility Studies or Basic Project.

This detailed characterization is fundamental to support erosion studies downstream of each type of spillway in hydraulic models: basins in the case of hydraulic jump and plunge pools in the case of ski jumps, as will be seen in this book.

1.3 Spillway and design flood

The spillway is the structure of the hydroelectric power plant destined to extravasate the exceptional floods in order to protect the dam against overtopping. This is especially important in the case of run of river power plants that do not have accumulation reservoirs.

In other words, it is the structure used to avoid exceeding the maximum water level of the reservoir. The excess is normally carried back to the river safely by an artificial chute. It should be noted that this cause is responsible for 23% of dam ruptures (USBR, Jansen, 1983).

For the selection of the inflow design flood, it is recommended to check DSD (1974) and ICOLD Bulletin 58 (1987), among others. It is also recommended to consult Afshar and Marino (1990). In Brazil, the selection of the flood is made according to the criteria of Eletrobras (1987), and the project of the spillway should be elaborated by viewing the "Criteria of Civil Project of Hydroelectric Power Plants" of Eletrobras/ CBDB (2003). This document, in Portuguese, is available on the internet. It specifies in its section 3.4.1:

- **for dams with a height > 30 m** or whose collapse involves the risk of loss of human life (existence of permanent dwellings downstream), the design outflow of the extravasating organs shall be the maximum probable flood (CMP);
- **for dams with a height < 30 m** or with a reservoir with a volume of $< 50 \times 10^6$ m^3 and where there is no risk of loss of human life (no permanent dwellings downstream), the design flood shall be defined by means of a risk analysis, with a minimum recurrence of 1.000 years.

The case of the Mascarenhas de Moraes HPP

On the review of the design floods, it was considered pertinent to show the case of the complementary spillway of the Mascarenhas de Moraes HPP, 478 MW, owned by CPFL (Figure 1.10), built in the Grande river, between the states of Minas Gerais (MG) and São Paulo (SP) from 1953 to 1957 between the Furnas HPP upstream and the Estreito HPP downstream.

The plant had a spillway with 11 spans with radial gates, 12,20 m wide by 10,67 m high, and maximum discharge capacity of 10.400 m^3/s.

In 1973, FURNAS acquired the plant and conducted a series of studies to establish new flood control procedures to protect the powerhouse during rainy periods.

In 1997, FURNAS reviewed flood studies by the method of maximum probable precipitation (PMP) using the SSARR model (Streamflow Synthesis and Reservoir Regulation) from the Corps of Engineers (USA). It was concluded that the maximum discharge capacity of this

Figure 1.10 Mascarenhas de Moraes HPP.
Source: (Minas Gerais/São Paulo, Brazil) (CBDB, 2002).

spillway should be 13.656 m³/s, which necessitated an additional spillway of 3.300 m³/s – 40 years later.

This new spillway on the left abutment, 12 m wide by 19,12 m high (Figures 1.11 and 1.11A), was built from November 1998 to February 2002. The chute with a ski jump at the end is 306 m long.

Figure 1.11 Mascarenhas de Moraes HPP. Additional spillway on the left abutment.
Source: (Minas Gerais/São Paulo, Brazil).

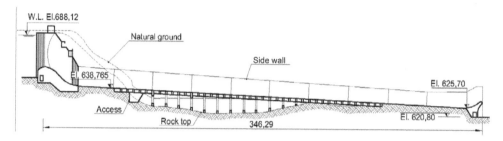

Figure 1.11A Mascarenhas de Moraes HPP, also known as Peixoto. Additional chute spillway. Longitudinal profile.

Source: (CBDB/LBS, 2002).

For details see the publication CBDB/LBS – Large Brazilian Spillways (2002), from which Table 1.1 was extracted.

It is important to mention the case of the Palagnedra dam, on the Melezza river in southern Switzerland, built between 1950 and 1953; the spillway was remodeled (Vischer and Trucco, 1985).

The dam has an arc-gravity cross-section, with a height of 71 m and a crest width of 120 m. The original spillway was uncontrolled to 450 m³/s, with 13 openings of 5,0 × 3,0 m each. The access road to the city of Palagnedra passes by the crest of the dam (Figures 1.12A and 1.12D)

In August 1978, an extraordinary flood occurred that caused a great erosion in the right bank that caused a race of debris to the dam. The race also carried a lot of vegetation, tree trunks, and a lot of debris transported by the flood wave partially obstructed the openings of the spillway and caused overtopping of the dam which resisted well (Figure 1.13B), but the highway was interrupted.

Immediately the owner cleared the highway and contracted hydrological studies to review the design flood. The new flood was set at 2.200 m³/s for 1.000 years of recurrence time, including a 20% gap to address the debris issue.

The model studies were conducted by the Laboratory of Hydraulic, Hydrology and Glaciology of the Federal Institute of Technology of Zurich. The water level of the reservoir was fixed at elevation 492,25 m, with freeboard of 1,75m. Figure 1.12D shows the refurbished project.

To complete this section, it is worth remembering the accident with the dams of the Pardo river, in the Northwest Region of the State of São Paulo, shown in the following profile (Figure 1.13): Armando Sales de Oliveira (Figure 1.14), Euclides da Cunha (Figure 1.15) and Caconde (Figure 1.16). After an exceptional flood in January 1977, the Caconde dam, constructed between 1958 and 1966, was unharmed because its reservoir (area = 31 km², volume = 540 × 10⁶ m³) cushioned the flood volume. The Euclides da Cunha and Armando Salles de Oliveira dams, built between 1958 and 1960, were overtopped and broken (BCOLD, 1982).

After this accident, the hydrological studies of floods were reviewed, and it was decided to implant two additional spillways: one in Euclides da Cunha and another in Caconde.

Table 1.1 Main Brazilian spillways (CBDB, 2002)

Name	Q m³/s	Q m³/s/m	ΔH m	Gates Tipo	No	L (m)	H (m)	Dissipator
Água Vermelha	20.000	165,40	~ 52	R	8	15,0	19,6	FB and plunge pool
Aimorés	15.000	92,59	~ 21	RI	10	13,5	15,3	BRH
Balbina	5.840	143,14		R	4	13,5	15,3	BRH
Barra Bonita	4.200	72,48	31	R	5	12,5	9,4	RB
Boa Esperança	8.633	92,83	32	R	6	13,0	12,7	FB
Camargos	2.030	26,71	27	R	6	11,0	6,0	BRH
Cana Brava	17.800	197,80		R	6	15,0	20,0	BRH
Canoas I	5.850	82,98	30	R	4	15,0	13,8	BRH
Canoas II	5.650	80,14	25	R	4	15,0	13,95	BRH
Capivara	17.100	118		R	8	15,0	15,56	FB and deflectors
Cachoeira Dourada	11.710	59,74	33	R	10	16,0	11,0	BRH, rock basin
C. Dourada Auxiliar	7.210	36,79		R	10	16,0	8,0	BRH, rock basin
Chavantes	3.200	~ 65,3	~ 70	R	3	14,0	15,32	Chute, basin/blocks
Corumbá	6.800	~129,5	~ 52	R	4	13,0	17,1	FB and plunge pool
Coaracy Nunes*	12.000	~71	23,8	R	10	13,8	12,5	BRH
Dona Francisca	10.620	31,70	> 40	–	–	–	–	Steps; BRH
Emborcação	7.600	129,90	~ 85	R	4	12,0	18,77	FB and plunge pool
Estreito	13.000	155,00	65	R	6	11,5	16,6	FB and plunge pool
Foz do Areia	10.250	145,18	130	R	4	14,5	19,45	FB and plunge pool
Funil (RJ)	1.700		~ 88	R	1	11,5	16,5	Tunnel, BRH
Funil (RJ) Auxiliar	2.700		~ 88	R	2	13,5	14,2	Tunnel, BRH
Funil (MG)	7.356	147,12		R	4	12,5	15,0	BRH
Furnas	12.297	114,70	89,3	R	7	11,5	15,8	FB and plunge pool
Ibitinga	5.163	37	27	W	7	11,65	6,0	BRH
Ibitinga Auxiliar	2.139	81,85	31	W	3	8,65	4,75	Pool
Igarapava	14.300	145,18		R	6	13,5	18,15	BRH
Ilha Solteira	34.300	97,6	47	R	19	18,5	21,5	BRH
Irapé	7.950		107	R	3	11,0	20,0	Tunnel, FB/p. pool
Itá	29.964	92	230	R	6	18,0	21,86	Chute/FB/plunge pool
Itá Auxiliar	19.976	80	234	R	4	18,0	21,86	Chute/FB/plunge pool
Itaipu	62.200	85	185,67	R	14	20,0	21,34	Chute, FB
Itaparica	26.415	48	150,94	R	9	15,0	19,7	RB
Itapebi	20.915	> 80	168,13	R	6	17,4	20,0	SE
Itauba	8.130	> 83	147,82	R	3	15,0	20,2	SE
Itumbiara	16.270	> 75	180	R	6	15,0	18,85	FB and plunge pool
Nova Ponte	6.140	99,68		R	4	11,0	17,35	Chute/FB/plunge pool
Orós	4.400	24,44		–	–	–	–	FB and plunge pool
Passo Fundo	2.262	25		R	6	12,0	5,75	BRH
Paulo Afonso IV	10.000	86,21	101	R	8	11,5	19,6	Chute, SE
Pedra do Cavalo	12.000	130,43		R	5	15,0	18,0	Chute, FB
Peixoto (M. Moraes)	11.026	109,0	46	R	11	12,2	10,67	FB
Peixoto Auxiliar	3.593	130,65		R	2	12,0	19,12	FB
Porto Colômbia	16.000	98	16	R	9	15,0	15,4	BRH
Porto Primavera	52.800	216,7	18	R	16	14,96	22,86	BRH
Salto Caxias	49.600	215	60	R	14	16,5	20,0	FB
Salto Osório	15.000	167	> 46	R	5	15,3	20,77	Chute, FB
Salto Osório Auxiliar	12.000	167		R	4	15,3	20,77	Chute, FB
Salto Santiago	24.600	164	88	R	8	15,3	21,57	Chute, FB
São Simão	24.100	147,85	59	R	9	15,0	18,78	Chute, FB/p. pool
Segredo	16.000	160	48	R	6	14,0	21,0	Chute, FB/p. pool

(Continued)

Table 1.1 (Continued)

Name	Q m³/s	Q m³/s/m	ΔH m	Gates Tipo	No	L (m)	H (m)	Dissipator
Serra da Mesa	14.750	166		R	5	15,0	19,0	Rock chute
Sobradinho bottom	16.000	102,56	21	R	12	9,8	7,5	DF, RH
Sobradinho surface	6.855	107,11		R	4	9,8	7,5	BRH
Três Irmãos	9.000	150	46	R	4	15,0	19,0	BRH
Três Marias	8.700	145	51	R	7	11,0	13,7	FB and plunge pool
Tucuruí	110.000	207	49	R	23	20,0	21,0	FB and plunge pool
Volta Grande	16.580	> 90		R	10	15,0	11,4	BRH
Xingó	16.500	151,38	108	R	6	14,83	20,76	FB
Xingó Auxiliar	16.500	151,38		R	6	14,83	20,76	Chute, SE

R = radial gate; W = fixed-wheel gate; RB = roller bucket; FB and plunge pool = flip bucket; SE = ski jump;
BRH = hydraulic jump basin; DF = bottom outlet.
* Included by the author. Eletronorte reviewed the hydrological studies in 2014. This flood was reduced by 40%, from 12.000 m3/s to 7.213 m3/s, after 38 years of operation.

Figure 1.12 Palagnedra dam spillway. Switzerland.

Source: (Vischer and Trucco, 1985).

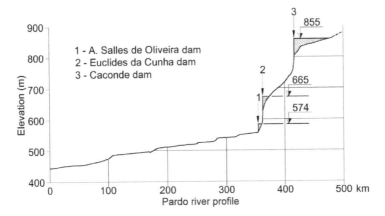

Figure 1.13 Pardo river profile. Brazil.

Rupture.   Recovered dam.

Figure 1.14 Euclides da Cunha dam.
Source: Images (BCOLD, 1982).

The original Euclides da Cunha spillway has a capacity of 2.040 m³/s. It was decided to install a spillway for a further 1.060 m³/s (52%). A morning glory spillway was adopted to take advantage of the old diversion tunnel, due to the lack of space to increase the discharge capacity of the existing spillway.

The project was supported by hydraulic model studies conducted at FCTH-USP. The final design resulted in a morning glory spillway with diameter of 27,10 m and a shaft with a diameter of 8,90 m. The maximum velocity in the tunnel is 36 m/s. Details can be found at Oliveira and Leme (1985).

The Caconde HPP had only one spillway, morning glory type, with a capacity of 718 m³/s. A new chute spillway for 900 m³/s was built on the left abutment between 1998 and 1992, as shown in Figure 1.16.

A   B

C

Figure 1.15 Armando Salles Oliveira dam.
Source: Images (Oliveira and Leme, 1985).

Figure 1.16 Caconde HPP
Source: (abrage.com.br).

1.4 Layout and choice of spillway

The layout and type of spillway depend on the general arrangement of the works of a hydroelectric power plant, which in turn depends on the topographical and geological characteristics of the site.

In general, the spillway has five distinct elements: approach channel, control structure, chute channel, energy dissipator and exit channel.

Several types can be adopted, as presented by Guinea (1973), Johnson (1973), Rudavsky (1976) and Rao (1982). Among others, these factors can be highlighted in summary:

Spillway in regard to the position of the plan in the arrangement of works

- can be incorporated into the main dam body; or
- be a separate structure of the dam such as side spillways on the abutments or side channel or chute spillways, or a morning glory spillway.

Spillway as to elevation position

- surface spillway, free or controlled by gates, allows lowering of the reservoir level to its crest.
- bottom spillway, controlled by gates, allows the total or partial emptying of the reservoir.

Spillway – profiles and cross-sections

- cross-sections with upright facing and sloping downward-facing Creager profiles; after the Creager profile, one can have a stilling basin or a spillway chute with a ski jump at the end; this profile is often stepped spillways, without control of gates;
- cross-sections of rounded crest with jet in free fall, or hole with ski jump (or cross jets) in the case of arc dams in narrow valleys;
- cross-sections in shaft-type morning glory spillway;
- cross-sections in orifice (bottom), in conduit (under a dam) or in tunnel;
- cross-sections in siphon, used when it is necessary to automatically spill an excess inflow to keep the reservoir water level at a specified elevation.

There are many classifications and particulars. In HDS (1965), the material is presented according to the following itemization:

- high overflow spillways;
- low overflow spillways;
- chute spillways;
- morning glory spillways, also known as "shaft spillways";
- auxiliary spillways or emergency spillways.

It was decided to present the theme of this book with emphasis on the topographical aspects of the site:

- **in a broad plain**, a spillway incorporated into the dam is adopted (e.g., Tucuruí, Balbina, Sobradinho, and many others that will be shown later); Figure 1.17 shows the example of the Santo Antônio Plant on the Madeira river, with two spillways incorporated into the dam.

Figure 1.17 Santo Antõnio HPP, Madeira river (Brazil-RO).

- **in a moderately embedded valley** one can adopt a spillway (Barra Grande, Campos Novos, Corumbá and others); Figure 1.18 shows the example of Itapebi, Jequitinhonha river (BA).

Figure 1.18 Itapebi HPP. Brazil.
Source: (www.itapebi.com.br; see Xavier, 2003).

- in a "V" valley, a lateral spillway (e.g., Hoover, Funil, Monjolinho, among others), a morning glory spillway (Monticello), shown in Figure 1.19, or a spillway above the dam/powerhouse, as in Karakaya, Figure 1.20 (Euphrates river, Turkey); Figure 1.21 shows the Chirkeysk Power Plant in Russia, (Grishin, 1982), an arc dam 236 m high with a lateral spillway followed by a tunnel in the left abutment.

The spillway of the Itaipu Plant deserves a special mention, Figure 1.22. Each case should be carefully studied.

The consolidated experience has shown that even observing the Project Criteria, these structures have presented cavitation and erosion problems downstream – see cases recorded in this book and in Main Brazilian Spillways (CBDB, 2002), among others.

The design is complex, and it is not enough to follow the hydraulic design guidelines. For large structures, it is always necessary to rely on models, bi- and three-dimensional, to test the operation of all the works and the structure itself.

Figure 1.19 Monticello dam. See section 2.4.

Source: Napa, CA-EUA.

Figure 1.20 Karakaya HPP, Turkey. Details in section 4.3.

Note: This solution, a spillway built into the dam above the powerhouse, is more expensive, but cleaner, since it minimizes the impact of the excavations at the abutments, i.e., they are drastically reduced when compared to the Itapebi solution (Figure 1.18) and with Barra Grande and Campos Novos solutions (Figures 2.18 and 2.20), where the impact on the abutments is visible.

Figure 1.21 Chirkeysk HPP, Russia. Lateral tunnel spillway.

Source: (wikipedia.com).

Figure 1.22 Itaipu HPP. Brazil.

1.5   Two notable cases: Theodore Roosevelt (USA) and Orós (Brazil)

The following are the cases of two examples of notable spillways, containing photos and typical cuts of these structures. One of them is the Theodore Roosevelt (Figures 1.23 and 1.24) spillway, which has been completely remodeled to continue operating safely. The other is the spillway of Orós dam (Figure 1.27, Brazil), which was undersized, and the dam was overtopped and breached. It was then rebuilt.

The first example is the project of the Theodore Roosevelt HPP, on the Salt river in the state of Arizona (USA). It was the highest masonry dam in the world, 85,30 m high, 220,4 m long, 4,9 m wide at the crest and 56,1 m at the base, built by USBR between 1903 and 1911. The slope of the dam had a slope of 0,05:1 and a downstream slope of 0,67:1. In addition, the reservoir provides both flood control and irrigation, which transformed part of the Arizona desert into a farm with very productive land, where the main crops are wheat, sorghum, pasture, alfalfa, barley and fruit.

Considering the good conditions of the rocky massif at the site (gabbros, dolomites and chert, a variety of fractured quartz), very little excavation was required. The joints were sealed with cement injection. The dam had no drainage gallery, and the forces of underpressure were not considered.

The pressures at the foot of the dam, upstream and downstream, were compared with compression tests on two cubic specimens extracted from the foundation. The results indicated that the rock had sufficient strength to withstand the stresses.

Construction was interrupted several times by floods. The worst of them, in 1905, destroyed the cofferdams. The original powerhouse had a capacity of 19,3 MW on several 25 Hz generators. In 1973, these generators were replaced by a 60 Hz generator with a capacity of 36 MW and a maximum head of 68,6 m. The plant has two penstocks of 3,0 m in diameter that join into one of 4,42 m before the unit.

The Theodore Roosevelt HPP has two spillways with unlined chutes (Figure 1.25), one on each abutment, for a total discharge of 4.248 m³/s. The spillways were originally without control – crest free. Figure 1.25 shows the two spillways operating.

In 1913, the crests were raised by 1,50 m to increase the volume of the reservoir. In 1923, segment gates were installed: 10 in the left spillway and 9 in the right

Figure 1.23 Theodore Roosevelt HPP, 1911.
Source: Wikipedia.

Figure 1.24 Theodore Roosevelt.
Source: (Kollgaard and Chadwick, 1988).

Figure 1.25 Spillways operating.
Source: Kollgaard and Chadwick (1988).

spillway, all with dimensions of $6,10 \times 4,80$ m. The total width is 122,50 m. The specific discharge rate is low, on the order of 34,70 $m^3/s/m$, considering the poor quality of the massif.

Throughout its useful life, no infiltration or structural problems were found. In 1907, 1940 and 1974 the stability analyses were repeated, and the conclusions were that the dam could withstand the dynamic loads from a magnitude 5,8 earthquake. The passage of the floods of 1915 and 1916 caused erosion of the channels of the spillways. The scours were repaired with concrete.

In 1935, small-scale studies were conducted at Colorado State University (CSU). The tests recommended lowering the crest by 1.5 m to restore the discharge capacity of 4.248 m^3/s.

In December 1978, considerable erosion damage was reported and repaired immediately. In the 1980s, the USBR conducted studies of alternatives that involved design modifications to fit the PMF (probable maximum flow) and to correct design deficiencies related to the safety of structures with respect to new studies of dynamic loads.

In eight years (1989 to 1996), the dam was modified by the USBR, being raised 23,50 m, and the section was transformed into a classic concrete-gravity arc cross-section, with the objective of solving the safety problems. The stored volume increased by 20%.

Two new spillways were constructed (Figure 1.26). A new bottom outlet has been installed. The powerhouse has been completely renovated. The recreation facilities have also been upgraded. The upstream road bridge is from 1990.

Figure 1.26 Rebuilt plant.
Source: Wikipedia.

The second example is the spillway of the Orós dam (Figure 1.27), in the Jaguaribe river in the state of Ceará, Brazil (CBDB/MBD, 2000; CBDB/LBS, 2002).

Figure 1.27 Orós, dam and spillway. Ceará, Brazil.
Source: (CBDB/MBD, 2000; CBDB/LBS, 2002).

The rockfill dam, with a top elevation of 209,00 m and a semicircular axis plant with a radius of 150 m, 54 m high and 620 m in length (see Figure 1.28), was designed and built by the DNOCS (National Department of Drought Works) from 1958 to 1961.

The cross-section (see Figure 1.29) has a compacted clay core (1H: 1 V) with a base of 100 m width, transitions of sandy gravel (1V: 2H) and rockfill (1V: 2,5H). The spillway was designed in 1962 and built between 1964 and 1966 (CBDB/LBS, 2002).

The spillway was designed for a discharge of 4.400 m^3/s (WL = 207,00 m), with a final capacity of 5.200 m^3/s (WL = 207,80 m). The net width of the crest is 146 m. The plant and the profile of the spillway are shown in Figures 1.30 and 1.31.

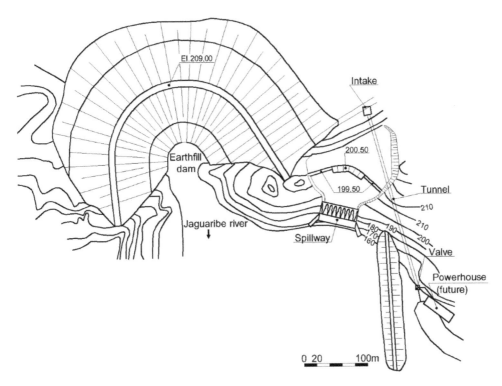

Figure 1.28 Orós dam. Arrangement of the works.
Source: (CBDB/MBD, 2000).

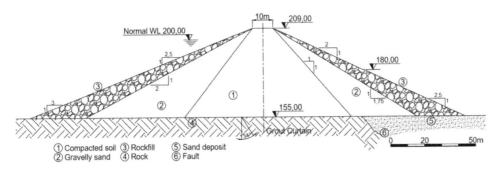

Figure 1.29 Orós dam cross-section.
Source: (CBDB/MBD, 2000).

According to CBDB/MBD, 2000, excavations of the dam were completed in October 1958 and were ready to receive the embankment. The schedule provided that the dam would be operative during the 1960 rainy season and the flood would pass through the spillway at elevation 200,00 m.

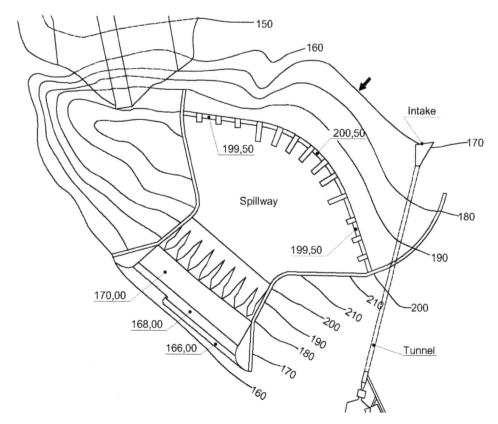

Figure 1.30 Orós spillway. Plant.
Source: (CBDB/LBS, 2002).

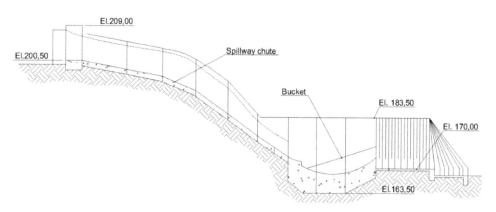

Figure 1.31 Orós spillway – profile.
Source: (CBDB/LBS, 2002).

The rains, which arrived late and weak in early 1960, intensified in March in such a way that they hampered the embankment. The height of the dam reached 183 m, 17 m below the sill of the spillway.

On March 12, a discharge of 400 m³/s passed through the adduction tunnel to the future powerhouse. The water level of the reservoir rose in nine days to the elevation 180,00 m. Additional precipitations raised the discharge to 2.000–2.250 m³/s. It was decided to raise the crest of the dam, working for 24 hours in the construction of the embankment.

The dam had not yet reached the 190,00 m elevation (Figure 1.32) when water began to crest on the night of March 25/26 (Figure 1.33). The head on the dam was 0,30 m.

It was judged to be more effective to control the damage by opening a breach in the land-fill to facilitate the passage of the peak of the flood, which opened the remainder of the breach to the foundation of the dam. At the breach, the head was 0,80 m. A total of 765×10^3 m³ of the landfill was removed from a total of $2,0 \times 10^6$ m³, opening a gap 200 m wide.

The volume of stored water, 730×10^6 m³, was reduced to 70×10^6 m³ in 24 hours. The peak of the flood after 12 hours was 9.600 m³/s. One hundred thousand people were evacuated from the Jaguaribe river valley by the army.

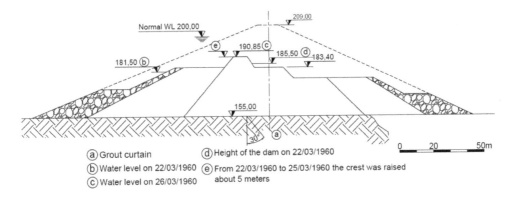

ⓐ Grout curtain
ⓑ Water level on 22/03/1960
ⓒ Water level on 26/03/1960
ⓓ Height of the dam on 22/03/1960
ⓔ From 22/03/1960 to 25/03/1960 the crest was raised about 5 meters

Figure 1.32 Orós dam – cross-section.
Source: (CBDB/MBD, 2000).

Figure 1.33 Orós. Overtopping of the dam
Source: (CBDB/MBD, 2000).

The CBDB/MBD (2000) cites that life losses were minimized but did not cite how many people lost their lives. It records only that the army evacuated 100.000 people.

As soon as it was flooded, the dam was rebuilt and inaugurated on January 11, 1961, with the spillway excavated without being lined with concrete (CBDB, 2011). The local rock is composed of schists of the Ceará series, with prominent foliated and extremely fractured quartzites. The spillway area was protected by a cofferdam.

According to the CBDB (2011), a federal authority, visiting the site at the time the reservoir was full, ordered the opening of the cofferdam. The flow over the excavation caused great erosion (Figure 1.33A). Still in accordance with the CBDB (2011) and CBDB/LBS (2002), the spillway project was done by Hidroesb in 1962 (Figure 1.33B and 1.33C), and the construction was done directly by DNOCS between 1963 and 1966.

It should be recorded that this case, constructing the spillway only after the construction of the dam is finished, is unusual. Even more so for a dam 54 m high, implanted in a densely populated valley.

Figure 1.33A View of the excavated area of the spillway after overtopping. Note that the construction of the spillway had already begun.
Source: (CBDB/LBS, 2002).

Figure 1.33B Orós dam. Spillway working.
Source: (CBDB/MBD, 2000).

Figure 1.33C Orós – Hydraulic model.
Source: Hidroesb-RJ (CBDB/LBS, 2002).

1.6 Ruptures of dams due to insufficient discharge capacity of the spillway

Jansen – USBR (1983), in the book titled *Dams and Public Safety* presented several cases of dam rupture. The main causes are foundation problems, seepage, erosion, embankment movement, liquefaction, concrete deterioration, spillways, outlets, demolition, sliding and induced earthquakes.

The main cases of rupture due to insufficient discharge capacity of the spillway are South Fork dam – USA (1889), Orós dam – Brazil (already presented, 1961), Panshet dam – India (1961), Sempor dam – Indonesia (1967) and Machhu II dam or Morbi disaster – India (1979). The last three disasters each caused thousands of victims.

On July 12, 1961, the nearly completed Panshet dam in the vicinity of Poona in India was overtopped and washed away (Figure 1.34). The released waters poured into the Khadakwasla Reservoir a few miles downstream, causing it to overflow and then break. Continuing down the Mutha river, the flood hit the city of Poona and wrought widespread destruction. Approximately 5.000 homes were either damaged or demolished.

The 51-meter high Panshet dam, under construction since 1957, was a zoned earthfill structure with a side spillway having a capacity of 487 m³/s. There were plans for future addition of crest gates to enlarge the reservoir. An outlet was located in a trench on the left abutment.

Figure 1.34 Panshet dam rupture. India (1961).

Failure of the embankment portions of 26-meters high masonry and earthfill Machhu II dam in the western state of Gujarat on Saturday, August 11, 1979, was caused by overtopping during flood (Figures 1.35 and 1.36). About 700 meters of earth embankment on the right side of the dam and 1.070 meters on the left side were washed away. Loss of human life during the ensuing flood may have been as high as 2.000 or more. Maximum inflow into reservoir was reported to have exceeded 14.000 m³/s, almost three times the peak design inflow of 5.600 m³/s. The spillway has 18 gates, 9,14 m long by 6,10 high. At the time of the disaster, 15 gates were fully open, 01 were 4,9 m open, 01 were 1,83 m open and 01 were 1,22 m open.

Figure 1.35 Machhu II dam rupture (1979).

Figure 1.36 Machhu II dam rupture.

The book includes also the case of the El Guapo dam (Figures 1.37 to 1.41), situated in the city of Barlovento, in Venezuela, 150 km from Caracas, whose objectives are: flood control, water supply (415.000 people) and irrigation. The dam was 60 m high and 524 m long, with a spillway to 102 m³/s.

The rupture occurred on December 16, 1999. The recovery of the dam ended in 2009, ten years after the break, and the main reconstruction services involved: 30.000 m³ CCV; 350.000 m³ CCR; 350.000 m³ of compacted landfill. It should be noted that the new flood of the spillway, 2.700 m³/s, is 27 times larger than the original. The following are photos of this accident.

Figure 1.37 El Guapo. Venezuela.
Source: (Prego Ing. Geotecnica).

Figure 1.38 Overtopping. Erosion.

Figure 1.39 Dam break.

Figure 1.40 Reconstruction of the dam.

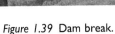

Figure 1.41 Dam rebuilt and spillway working.

Source: (Prego Ing. Geotecnica).

The case of the Opuha dam for irrigation in New Zealand (Figure 1.42), 50 m high, is also recorded that failed on May 2, 1997, during the construction, due to the occurrence of excess rainfall and lack of capacity of discharge of the discharge conduit (outlet) under the dam of 1,8 m in diameter (Lees and Thomson, 1997).

The maximum outflow estimated for the project was 1.500 m^3/s (PMF). There was a flood between 1.500 and 2.000 m^3/s. The period of construction of the dam, including reconstruction, was from 1996 to 1999.

The case of the Glashütte dam, 2002, Germany, a 9,0 m high retaining dam built between 1951 and 1953 on the Briesnitz river in the town of Glashütte, is also worth noting. The river Briesnitz is a tributary of the Müglitz river, which discharges into the Elbe near Dresden.

The dam is situated 2,0 km upstream from the city. The drainage area was 10,0 km^2. The spillway design flood was estimated at 5,0 m^3/s, and the width of the structure 5,5 m, with steps of 1,0 m. The dam was 10,0 m high and had a sewer with a capacity of 7,0 m^3/s, without control, where the river flowed. The total discharge capacity was 12,0 m^3/s.

A – Opuha – Downstream view of the breach.

B – Opuha – Upstream view of the breach.

C – Dam rebuilt.

Figure 1.42 Opuha dam, New Zealand.
Source: (Lees and Thomson, 1997).

In August 2002, heavy rains occurred in southern Saxony, causing a flood of 120 m³/s. The dam was overtopped and failed – Figures 1.43 and 1.44 (Bornschein, 2003). Figure 1.45 shows the damage to the city. The dam was recovered between 2005 and 2006 and enlarged between 2010 and 2013 (see Figure 1.46). The new spillway was designed to address this discharge.

Throughout this book, to illustrate the specific themes, several other examples of spillways will be presented.

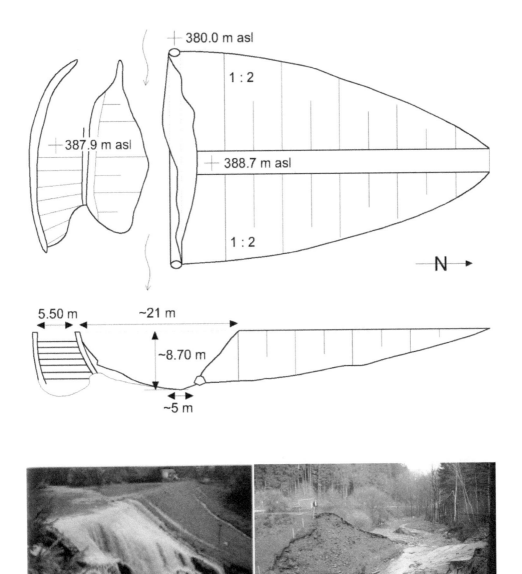

380.0 m asl

1 : 2

387.9 m asl

388.7 m asl

1 : 2

N →

5.50 m   ~21 m

~8.70 m

~5 m

Figure 1.43 Glashütte dam details. Overtopping photos and breach.
Source: Bornschein, A. (2003).

Figure 1.44 Glashütte dam breached. Stepped spillway damaged.

Source: Bornschein, A. (2003).

Figure 1.45 Glashütte city.

Source: Bornschein, A. (2003).

Figure 1.46 Glashütte dam rebuilt. Note morning glory spillway over the dam slope. Downstream view – note the spillway exit.

Source: (Courtesy of Bornschein, 2017; damfailures.org; commons wikimedia.org).

1.7   Risks of dam ruptures – submersion waves

The reservoirs of the hydroelectric plants of high and medium fall usually store a large volume of water. The main risk for the riverine population, as well as the surrounding property, is that of accidental and instantaneous release of this water, which will imply the formation of a submersion wave that will spread through the valley downstream, devastating everything.

The main phenomenon that causes a submersion wave is the rupture of the dam itself. This rupture can be progressive or instantaneous, partial or total, according to the type of dam. It can also result from a large landslide into the reservoir. Although rare, this was what happened in Vajont (Italy) on September 10, 1963. A huge volume (250×10^6 m^3) of Mount Toc material slid into the reservoir (168×10^6 m^3) of water, generating a big wave (h~100 m) that passed over the dam (which remained intact), submerging the city of Longarone and creating thousands of victims. According to the articles consulted, in the design phase, geological and geotechnical studies of stability of the steep slopes of the reservoir should have been done in more detail.

Table 1.2 provides a list of accidents with more than 50 casualties that have occurred since 1900. A more complete list is presented by USBR-Jansen (1983). In the ICOLD Bulletin 99, Dam Failures Statistical Analysis, are the detailed statistics of the types of dam ruptures up to 1995. It is observed that:

- ruptures by foundation, inadequate spillway (submersion, overtopping), poor construction (piping), uneven settlement and high pore pressure of earth dams are more frequent;
- structural ruptures of arch dams are rare.

The following figures, available on the internet, show photos of the accident.

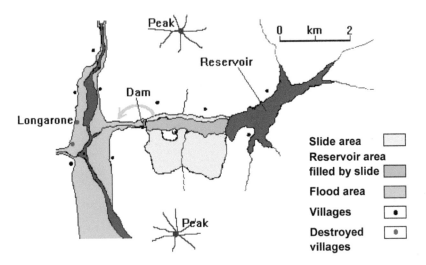

Figure 1.47 Accident of Vajont. Italy. Disaster map.

Source: (mannaismaya.adventure.com).

Figure 1.48 Accident of Vajont. Italy. A – Sliding. B – Overview after sliding.

Figure 1.49 Accident of Vajont. Italy. A – Downstream view of the dam. B – View of the dam for Longarone.

Source: (wikipedia.org; 3bmeteo.com; megaengenharia.blogpost.com).

Table 1.2 Partial list of accidents with more than 50 victims (Ginnochio, 2012)

Name/Country	Year/Circ. Rupture (1)	Year Constr	H(m)	L(m)	Res Vol(10^6m^3)	Type (2)	Nº of Victims
Bayless (USA)	1911/1ºE	1909	16	160	1,3	BGC	80
Bila Desna (Czech Republic)	1916/EI/1ºE	1915	18	240	0,4	BT	65
Tigra (India)	1917/C	1917	25	1.340	124	BGM	1.000
Gleno (Italy)	1923/1ºE	1923	35	225	5	BAM	600
S. Francis (USA)	1928/1ºE	1926	62	213	47	BGC	450
Alla Sella Zerbino (Italy)	1935/C	1924	16	70	10	BGC	100
Möhne (Germany)	1943/G	1913	40	–	134	BGM	1.200
Eder (Germany)	1943/G	1914	48	400	200	BGM	100?
Heiwaike (Japan)	1951/S	1949	22	82	0,2	BT	100
Malpasset (France)	1959/F	1954	60	222	48	BAM	423
Vega de Tera (Spain)	1959/1ºE	1955	33	270	7,3	BC	14
Khadakwasia (India)	1961/RM	1879	33	1.400	137	BGM	1.000
Panshet (India)	1961/SC	–	49	740	214	BT	1.000
Hyokiri (Korea)	1961/EI	1940	15	110	0,2	BT	139
Sempor (Indonesia)	1967/SC	–	60	228	56	BE	200
Nanaksagar (India)	1967/EI	1962	16	19.300	210	BT	100
Banqiao (China)	1975/S/T	1952	24,5	–	492	BT	>20.000
St. Thomas (Philippines)	1976/SC	–	43	–	–	–	80
Machu (India)	1979/FC/S	1972	26	3.900	101	BT	2.000
Gotvan (Iran)	1980/S	1977	22	710	–	–	200
Kantale (Sri Lanka)	1986/EI	1869	27	2.500	135	BT	127
Shakidor (Pakistan)	2005	2003	–	148	–	–	70
Situ Gintung (Indonesia)	2009/S	1933	16	–	2	BT	100

Notes:
(1) Rupture circumstance: 1º E = first filling; EI = piping; C = flood; G = war; S = submersion; SC = submersion during construction; F = foundation; RM = upstream rupture; FC = failure of the gates; T = typhoon.
(2) Type: BGC = concrete gravity dam; BGM = masonry gravity dam; BA = arch dam; BAM = multiple arches dam; BC = buttress dam; BT = earthfill dam; BR = rockfill dam.

1.8 Case of rupture by earthquake

The Shihkang dam, on the Da-Jia river in the central part of the island of Taiwan, provides the water supply in Chi-Chi (Jiji), Nantou Province. The height of the dam is 21,4 m and the length 357 m. The spillway capacity is 8.000 m³/s, with 18 gates 12,8 m wide × 8,0 m high.

The rupture was caused by an earthquake of 7,7 degrees on the Richter scale on September 9, 1999, generated by the movement of a major fault downstream of the dam. However, a secondary fault, more or less perpendicular to the first one, crossed the spillway along the right bank. Note the level difference, of the order of 10 m, between the portion along the margin and the remnant of the dam. There were 2.425 deaths and 11.305 injuries.

A

B

Figure 1.50 Rupture of the Shihkang dam, Taiwan.

Source: (researchgate.net; hss.dk)

1.9 Operation and maintenance

Every spillway must be operated according to the plan of operation of the gates (POG), which contemplates the whole range of flows to be discharged throughout its useful life, in normal and emergency situations.

The POG is elaborated in the basic/executive project phase. For large spillways, the various possible gate operating configurations are tested in hydraulic models, as already mentioned.

These plans must be followed by the operation team of the plant, in view of the safety of the powerplant. Before starting the operation, the operator team must be trained. The rules should be clear and well defined for the various levels of operational decision making and should cover all normal, exceptional and emergency operating situations.

The following conditions must be observed, among others:

- in large spillways, it is important to define the ranges of discharge variations to be discharged, in order to minimize the rapid variation of the water level downstream, which can cause downstream accidents (whether with riverine populations or with fishing commercial vessels, etc.); in emergency cases of sudden change in discharge the population likely to be affected must, in some way, be warned;
- minimize, whenever possible, the asymmetrical maneuvering of gates;
- consider the possibility of gates jamming, or in maintenance, and defining procedures for such situations;
- predict the possibilities of maintenance in the spillway considering the space of time, according to the hydrological regime available for these activities;
- consider the alternative that it might be necessary to build a cofferdam; and
- consider all assumptions aimed at increasing the operational flexibility of the structure.

Aspects relating to the maintenance of gates, maneuvering equipment and of the energy supply systems, as well as the risks involved in the event of gates becoming unavailable and other problems occurring during the flood season, are presented in Chapter 9.

This subject, of paramount importance for the security of the enterprise, is often forgotten. The experience in the operation of these structures records facts that deserve attention, as described later.

In the Jaguara HPP, which was completed in 1971, the first gate of the spillway to operate, the one on the left, should be the last one according to the POG, as shown by Magela and Brito (1991).

This fact occurred because this gate was the first to be ready, which shows that, for some reason, the assembly planning was not in line with what was established in the POG.

In the first year of operation, where the discharges were concentrated in bay 1 (on the left side of the water outlet/powerhouse), erosion evolved rapidly, reaching 23,00 m depth from the rocky surface, in a well-defined pit that reached the elevation 493,00 m (Figures 1.51 and 6.7 to 6.9).

In 1973, with the most distributed flow between the gates, erosion was generalized, with a maximum depth of 19,00 m at elevation 497,00 m, with loose blocks evidently covering the bottom of the previous erosion (Figure 6.9B).

In 1975 and 1977, topobathymetric surveys showed that erosion had reached 26,00 m at its lowest point, at elevation 490,00 m (Figures 6.9C and D), with the pit having the same shape.

In 1978, the erosion reached the maximum depth of 31,00 m, at the elevation 485,00 m (Figure 6.9E).

The last survey carried out in 1987 reveals that erosion remained, with 31.00 m from the rocky surface of the pre-excavation at elevation 516,00 m (Figure 6.9F), seeming to indicate a tendency of stabilization around this depth for the discharges so far spilled (Figures 5.4 to 5.10).

But as presented in CBDB (2002), the plunge pool was reshaped to solve the environmental problem of fish getting trapped in the plunge pool. An additional excavation was performed by CEMIG in 2001 connecting the plunge pool with the tailwater channel, as shown in Figure 5.9.

Figure 1.51 Scours downstream of Jaguara spillway. Brazil – Opening 01.

Source: Magela and Brito (1991).

In Kariba, the minimum discharges corresponded to extremely low downstream water levels (Hartung, 1973). This situation was more critical than the corresponding one for the project discharge. In this case, the POG contemplated this issue. Erosion downstream in 1982 already reached 90 m with respect to WL max. of upstream, elevation 400,00, as reported by (CBDB, 2002). Regarding this project, see also Anderson (1960), Hartung (1973) and Whittaker and Schleiss (1984).

As quoted in Chapter 6, the Zambezi River Authority has decided to contract the plunge pool to continue safe operation of the plant (see Figures 5.16 to 5.19). Projects in the process of contracting (October 2012) will involve additional excavation to alter the shape of the downstream slope of the erosion crater (plunge pool), making it smoother, as well as spillway modernization projects in separate contracts, according to Noret (2012) and Bollaert (2012).

In Tucuruí, during the drought of 1986, a special criterion of spillway operation was adopted, not foreseen in the POG, in order to improve the quality standards of downstream water consumed by the riverside population over 10 km. It was found that, by changing the spillway operating rule in this period, that aspect could be improved. The special scheme assembled sought to maintain the ratio of turbinate/total discharge defluent around 0,55. This was intended to cause an adequate mixing of the discharge from the spillway, with a high oxygen content, and the discharge from the water intake, with a low oxygen content, in order to improve water quality, minimizing the critical environmental effects both downstream and in the reservoir.

In Grand Coulee (see Figures 5.53 and 5.54), it was found that the spillages increased the nitrogen content in the downstream water, affecting the ichthyofauna, as reported by Jabara and Legas (1986).

These last two examples do not appear to have any interference with the downstream erosion of spillways, but special schemes of operation not provided for in the POG and designed to address environmental issues should adequately consider such interference to eliminate undesirable surprises.

The environmental impacts mentioned call attention to the fact that this variable must be considered. Other effects of the spills, such as wave action on the margins or navigation on the downstream stretch, for example, in specific cases may be associated with operational aspects and should be considered in the POG.

After commissioning, the performance of the structures must be monitored systematically, which is of fundamental importance not only to guide the possible maintenance/recovery of some damage but also to provide data that allow the subsequent assessment of the studies carried out in the phase of project. Scheduled maintenance of equipment and operation within specified ranges shall be strictly observed.

Mention is made of the accident at the Sayano-Shushensk plant (2009), where the turbine 2 exploded (Figures 1.52 to 1.54). In the accident, 75 people lost their lives. The research in articles available on the internet show a series of failures in the design of civil structures and equipment, as well as in the operation and maintenance of the plant. For details, it is recommended to consult Leyland (2010), Allen (2010) and Wenselowski (2106). According to these authors, such errors could have been avoided. It is worth highlighting the fact that an additional spillway in the right abutment was designed in the refurbishment of the plant (Figure 1.55).

Figure 1.52 Sayano-Shushensk HPP – original.
Source: (commons.wikimedia.org).

Figure 1.53 The accident.
Source: (Leyland, 2010).

Figure 1.54 Detail of turbine 2.
Source: (Leyland, 2010).

Figure 1.55 Sayano-Shushensk HPP reformed. Note the additional spillway on the right bank.
Source: (photo-day.ru).

Chapter 2

Types of spillways

2.1 Classification of the spillways

Spillways are normally "classified" according to their most important aspect:

- the manner of their control;
- the flow of their discharge channel;
- any other details, for example crest alignment.

As for control, spillways are routinely classified as "free" or "controlled by gates."

As for the discharge channel of the flow spilling, there are several types, quite different from each other. This book adopts a classification similar to that presented in the HDS (1965) and in the DSD (1973/1974):

- ogee spillway;
- side spillway;
- morning glory spillway;
- tunnel spillway;
- chute spillway;
- culvert spillway or bottom outlet;
- orifice spillway;
- labyrinth spillway;
- siphon spillway; and
- stepped spillway.

It is possible to include the non-ogee spillway. As will be presented at the end of this section, this spillway is composed of a bottom slab, piers and gates and is routinely used.

This classification is based on the selected arrangement, which depends on site boundary conditions, topography and rock mass conditions (geology).

In most large dams, the spillways are controlled, equipped with gates to allow flexible operation. This allows, with variation of the water level of the reservoir, the control of floods and the benefit of greater storage.

The cost of the gates increases with the magnitude of the floods, i.e., with the area of spilling. Inappropriate operations or improper operation of the gates is the point of greatest concern, which can result in dam overtopping.

Vischer and Hager (1998) approach the theme in a simpler way: frontal spillway, side spillway and shaft spillway.

According to these authors, other types, such as the "labyrinth" spillway, use a frontal spillway, but with the crest consisting of successive triangles or trapezoids in plan. Besides these, they cite also the spillway in "orifice."

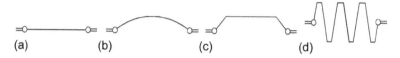

Figure 2.1A Sketching of spillway plants. Legend: (a) pattern crest – straight; (b) curved crest; (c) polygonal crest; (d) labyrinth crest.

2.2 Ogee spillway

This spillway has a prominent control apron which is called "ogive" (ogee), in profile. The upper shape of the ogee is normally made to conform to the profile of the aerated nappe over sharp-crested weir, as shown in Figure 2.1A.

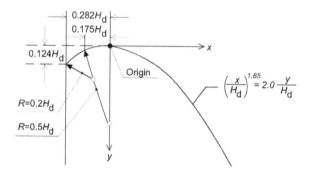

Figure 2.1B Ogee standard crest.

Source: HDC (1959).

The profile of this nappe has the classical equation recorded in the following illustration, Figure 2.1B, subject that will be discussed in more detail in Chapter 3.

This type of spillway is the main focus of this book and will be presented in detail in Chapter 3, through the example of the Tucuruí HPP spillway.

Several other examples of this type of spillway will be presented, including those in which the powerhouse is incorporated into the body of the structure.

In such cases, the spillway is the roof of the powerhouse. Examples are the Fengtan HPP (800 MW) in China (Figure 2.1C), which will be presented in Chapter 5, and the Karakaya HPP in Turkey (Figure 1.20), which will be presented in Chapter 4.

Figure 2.1C Fengtan HPP (400 MW). River Yoshui, Zhangjiabe, Hunan Province (China).
Source: (WP&DC, April 1989).

2.3 Side spillway

A side spillway can be used when the topographic characteristics of the site imply a solution for the arrangement of the dam works with the spillway not incorporated into the axis of the main dam of the river.

A schematic example is shown in Figure 2.2. In general, it consists of an approach channel/side channel crest, a side channel trough, a discharge channel (chute), an energy dissipator and/or a terminal structure (stilling basin).

The discharge characteristics are shown schematically in Figure 2.3. Examples of hydro-electric plants in which this type of spillway is adopted will be shown later.

The theory of flow in a side spillway was developed by Hinds (1926) and is based on the law of conservation of linear momentum. It is assumed that the energy of the flow on the crest is dissipated through mixing with the flow in the channel and that the only force producing longitudinal movement in the lateral channel results from the falling of the surface of the water along the axis of the lateral channel.

It is also assumed that the friction resistance of the channel is small and can be neglected without affecting the accuracy of the calculations. This theory has been proven in hydraulics models and prototype experience.

The main hydraulic aspects are:

- the crest is often free, since the topography of the site allows it;
- the shape of the crest is the same as for the normal sill weir;
- the crest is sized by trial and error to prevent the maximum surface of the water from "flooding" the free crest;
- the crest is designed to provide a supercritical flow in the lateral channel for all discharges and to allow the dissipation of all the energy discharged, producing a uniform flow in the chute;
- the side channel cross-section for the first attempt is designed to imply a uniform reduction of the upstream velocity to downstream.

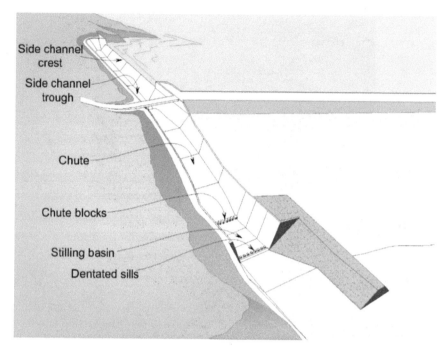

Figure 2.2 Side spillway. Typical arrangement.

Source: (DSD, 1960/1973/1974).

The waterline profile in the lateral channel can be estimated by Equation 2.1.

$$\Delta Y = \frac{Q_2(V_1 + V_2)}{g(Q_1 + Q_2)} \left[(V_2 - V_1) + \frac{V_1(Q_2 - Q_1)}{Q_2} \right] \tag{2.1}$$

where:
ΔY = change in the water level between two sections spaced of ΔX;
Q_1 and V_1 = discharge and velocity in the upstream section;
Q_2 and V_2 = discharge and velocity in the downstream section.

The calculation starts from a known downstream water level, equal to the critical depth (Yc), and proceeds to upstream in increments by trial and error, calculating the waterline along the lateral channel. If the calculated waterline profile invades the lateral channel capacity, the chute geometry or the crest, or both, must be adjusted. Once a geometry is determined for the channel and for the crest, the chute downstream must be tilted to provide a downstream supercritical flow.

In DSD (1960/1973/1974) there is a spreadsheet template for the calculations of this type of spillway. The structure needs to be tested in a reduced model because of the complexity of the lateral channel flow. Figures 2.4 and 2.4A show the example of the Monjolinho HPP. Figures 2.5A and 2.5B show the example of the Hoover dam.

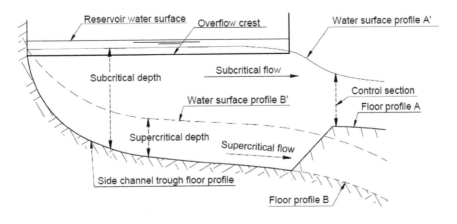

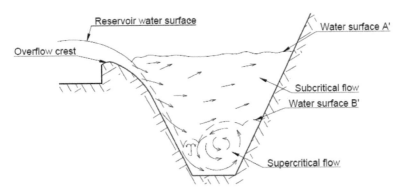

Side channel profile

<figure>
Figure 2.3 Side spillway. Flow characteristics.
Source: (DSD, 1960/1973/1974; Zipparro and Hasen, 1993).
</figure>

Figure 2.4 Monjolinho HPP. Side spillway on the right bank. $L = 210$ m; $Q = 6.755$ m^3/s.

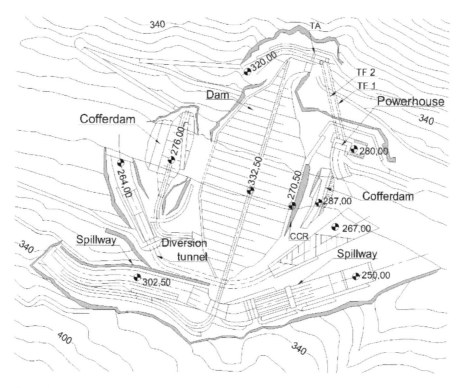

Figure 2.4A Monjolinho HPP, plant (Desenvix).

The Hoover dam on the Colorado river (Nevada/Arizona) has two side spillways followed by a tunnel, one on each bank, as shown in Figure 2.5A. The total discharge capacity is 11.000 m³/s.

Figure 2.5A Hoover HPP, Colorado river (Arizona/Nevada). Left side spillway operating (Arizona side).

Source: (www.dreamstime.com).

Figure 2.5B Hoover HPP. Detail of the side spillway of the left bank.

Source: (www.dreamstime.com).

2.4 Morning glory spillway

The morning glory spillway, also known as the shaft spillway, utilizes a crest circular in plan. The outflow is conveyed by a vertical or sloping shaft to a horizontal tunnel to downstream at approximately streambed elevation, as can be seen in the following figures. For details see HDC, Chart 140–1 (1955/1959), HDS (1965) and DSD (1973/1974).

According to HDS (1965), these spillways have been designed, with discharges varying from 70 to 1.500 m³/s, in places where the tunnel used in the phase of river diversion can be used as part of the horizontal discharge tunnel, and in places where the work arrangement includes a relatively low earth dam, as is the case of the PCH Colino in the Colino river, Bahia, Brazil (Figure 2.6). Figure 2.7 shows an illustrative cross-section of this type of spillway (USBR DSD, 1974).

Figure 2.6 Colino Small Hydro (Colino river, Bahia, Brazil). Spillway capacity Q = 143 m³/s.

It is also worth mentioning the Monticello dam spillway in the city of Napa, California. The dam, 93 m high, was built between 1953 and 1957 (Figure 2.8). It created lake Berryessa. The morning glory is 22 m in diameter on the surface and 8.5 m at the exit of the tunnel and has a capacity of 1.370 m³/s.

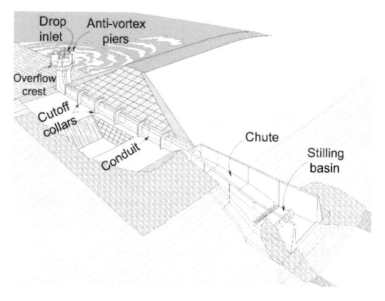

Figure 2.7 Morning glory spillway. Illustrative sketch.
Source: (Figure 240 USBR DSD, 1974).

Monticello dam (H = 93m).   Tunnel exit. D = 8,5 m.

Figure 2.8 Morning glory spillway of Monticello dam; $D = 22$ m. $Q = 1.370$ m^3/s.
Source: (Napa, CA-USA).

Another example is the morning glory spillway of the Harriman Power Plant, completed in 1924 (Figure 2.8A), of the New England Power Company, on the Deerfield river in the town of Whitingham, Vermont (USA). The earth dam is 55 m high and 381 m length. The morning glory has a diameter of 48,8 m, with piers and maximum flow capacity of 29.733 m³/s.

Figure 2.8A Morning glory spillway, Harriman HPP, Windham County, Vermont, USA.

2.5 Tunnel spillway

Tunnel spillways are used in places where there is no viability for the adoption of a chute spillway. Each site is unique, and as stated earlier, the solution is strongly conditioned by the topographical and geological conditions of each site.

According to the HDS (1965), tunnel spillways are routinely designed to work partially filled, with supercritical flow along the entire length. The ratio of the flow area to the design flood to the total tunnel area is 75%.

In several projects, the downstream cross-sections of diversion tunnels have been used as tunnel spillways, such as at Hoover HPP, Colorado river (Nevada/Arizona, USA), shown in Figures 2.9 to 2.11, as mentioned by Kollgaard and Chadwick (1988). In this plant, in each abutment, the tunnel is preceded by a side channel spillway.

Typically, the solution consists of an entrance structure, a sloping chute, the tunnel and a flip bucket at the end. The vertical curves in these tunnels are susceptible to damage by cavitation or erosion in projects with high falls (like Glen Canyon and Serre-Ponçon in Chapter 7).

The few available field information on the high velocity flows in these vertical curves and the respective pressures have led engineers to be more conservative in establishing the radius–diameter tunnel relationships (R/D – 7 to 10). Construction joints in the vicinity of the bends should be carefully treated to eliminate offsets and prevent pressure being transmitted through the joints.

It is worth mentioning the solution of Irapé HPP, in Jequitinhonha river, Minas Gerais state, Brazil, shown in the following illustrations (Figures 2.12A–D and 2.13), a rockfill dam 208 m high owned by CEMIG.

It should be noted that the lower tunnels were designed to guarantee the safety of the dam still under construction when the river is closed to start filling the reservoir; this operation was planned for when the embankment reached the elevation 475,00 m.

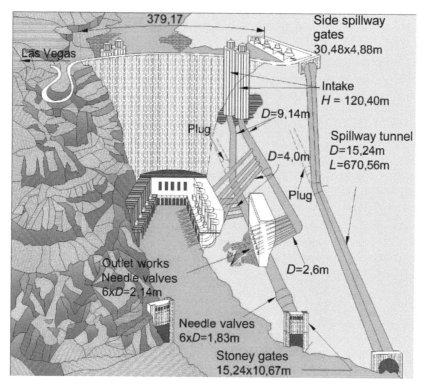

Figure 2.9 Hoover dam. Sketch of the works on the left abutment. H_{max} = 221,40 m. Discharge of the tunnels = 11.325 m³/s.

Figure 2.10 Hoover dam (1998). Observe the side spillways, Q_{max} = 11.325 m³/s.

Figure 2.11 - Hoover dam. Downstream view.

Figure 2.12A Irapé dam. General view.

Figure 2.12B Irapé HPP. Dam side view.

Figure 2.12C Irapé HPP. Detail of the intake and spillways structures on the left bank.

(D) Downstream tunnel exit.

Figure 2.12D Irapé HPP. Spillway in tunnel.
Source: (Magela 2015a).

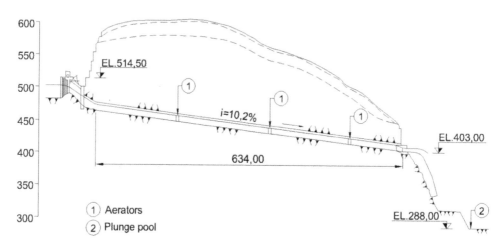

Figure 2.13 Irapé HPP. Spillway profile.
Source: (CBDB, 2009).

Spillway:

$Q = 4.000 \text{ m}^3/\text{s}$;
Two gates 11 m × 15 m;
Elevation of sill 491,00 m;
Tunnel chute, free flow; length = 626 m;
Cross-section 11 m (*L*) × 15 m (*H*), transition to 10 m (*L*) × 11,4 m (*H*).

Intermediate spillway:

$Q = 2.000 \text{ m}^3/\text{s}$;
Two gates 7 m × 9,4 m;

Elevation of sill 450,00 m;
Tunnel chute, free flow; length = 622 m;
Cross-section 17 m $(L) \times 15$ m (H), transition to 12 m $(L) \times 11,4$ m (H).

2.6 Chute spillway

A chute spillway consists of a low ogive followed by a sloped chute leading the flow downstream of the dam, as shown in Figures 2.14 and 2.15A–B – Itaipu HPP, Paraná river, Brazil/Paraguay. The spillway has 14 gates, 20,00 m $\times$ 21,34 m, and maximum discharge capacity of 72.000 m^3/s.

Figure 2.14 Itaipu HPP. Chute spillway.
Source: (CBDB, Highlights, 2006).

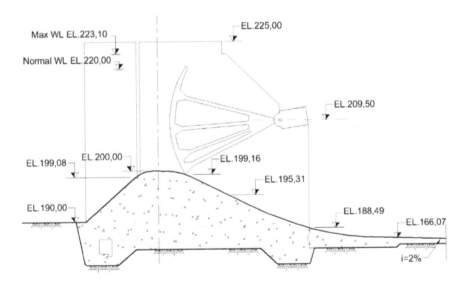

Figure 2.15A Itaipu spillway. Crest detail.
Source: (CBDB, 2002; ICOLD, Q.50, R.14, 1979).

The design of a chute spillway involves two problems associated with the supercritical flow: transverse waves of shock (of interference and of translation), as a function of the

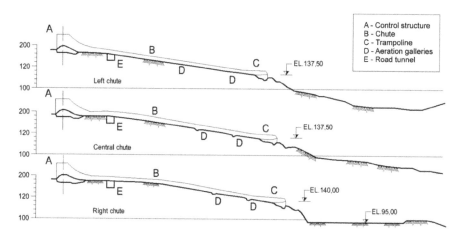

Figure 2.15B Itaipu spillway. Left, central and right chutes.
Source: (CBDB, 2002; ICOLD, Q.50, R.14, 1979).

curvature of the walls at the entrance and of the piers, and aeration along the chute (Figure 2.16). These waves cause a disturbance in the flow in the inlet that propagates downstream and will normally require a raising of the side walls of the chute, since the water tends to "accumulate" at the points where the waves meet the walls (A, B, C and D).

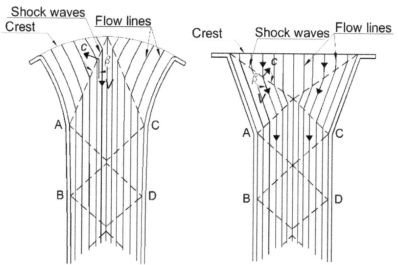

Figure 2.16 Shockwaves at the entrance of a chute spillway.
Source: (Novak et al., 2007).

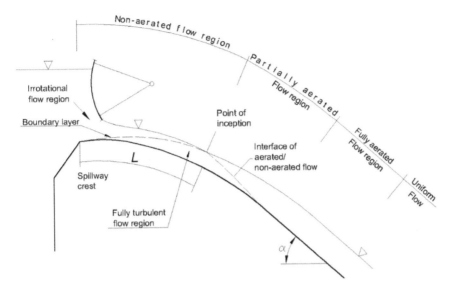

Figure 2.17 Development of the flow in a chute spillway.
Source: (Novak et al., 2007).

The detail of the treatment of shock waves can be found in Chow (1959) and Novak et al. (2007), who recommend that the spillway be projected to a wet depth/perimeter ratio > 0,1 for maximum discharge and to accept the propagation of waves for low flows.

Aeration details in Figure 2.17 show the most important aspect of supercritical runoff. It is beneficial to avoid cavitation, and to dissipate energy more causes an increase in the depth of the flow and therefore requires increasing the height of the side walls. This subject will be discussed in Chapter 8.

In summary, the profile of the waterline along the profile can be estimated with previously exposed. The freeboard and dimension of the walls should be checked in the hydraulic model studies. The following is a series of illustrations showing the various Brazilian projects where this type of solution was adopted.

Barra Grande HPP, Pelotas river (Rio Grande do Sul/Santa Catarina States, Brazil)

Chute spillway on left abutment in ski jump. Maximum flow capacity of 23.840 m3/s. Six gates 15 m wide by 20 m high (Figures 2.18 and 2.19).

The concrete chute is 274 m long with two aerator steps, to prevent cavitation of the concrete. Along the structure was provided a drainage gallery.

The local geology is dominated by the basalts of the Serra Geral Formation, with sub-horizontal distribution gradually dipping to the west.

Figure 2.18 Barra Grande HPP. Chute spillway/plunge pool.
Source: (CBDB, Highlights, 2006).

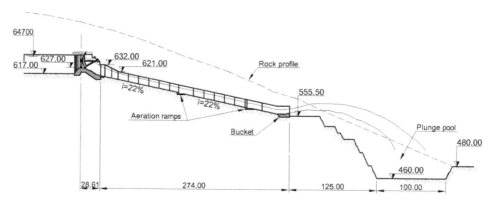

Figure 2.19 Barra Grande HPP. Chute spillway with ski jump/plunge pool.
Source: (CBDB, 2009).

Campos Novos HPP, Canoas river (Santa Catarina State, Brazil)

Chute spillway on the right abutment. Maximum flow capacity of 18.300 m³/s. Four segment gates 17,40 m wide and 20 m high.

Figure 2.20 Campos Novos HPP. Chute spillway.
Source: (CBDB, 2009).

The concrete chute is 94 m long; the energy dissipation is in a ski jump. Along the structure was provided drainage gallery (see Figures 2.20 and 2.21).

The local geology is dominated by the basalts of the Serra Geral Formation, with subhorizontal distribution gradually dipping to the northwest.

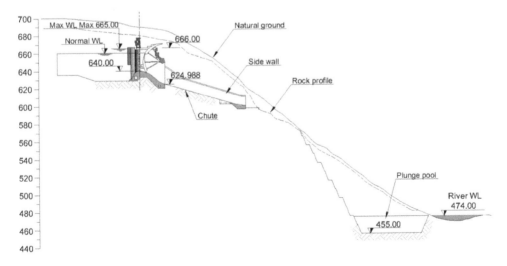

Figure 2.21 Campos Novos HPP. Chute spillway/ski jump.
Source: (CBDB, MBD, 2009).

Corumbá HPP, Corumbá river (GO, Brazil)

Chute spillway on the right abutment. Maximum discharge capacity of $Q = 6.800$ m³/s. Four segment gates of 13 m × 17 m.

The energy dissipation is in a ski jump. The spillway is founded on sound rock composed of chlorite – schist.

The concrete chute is 64 m wide and 150 m long, in slab with 0,70 m thick anchored in the rock. Along the structure was provided a drainage gallery (see Figures 2.22 and 2.23).

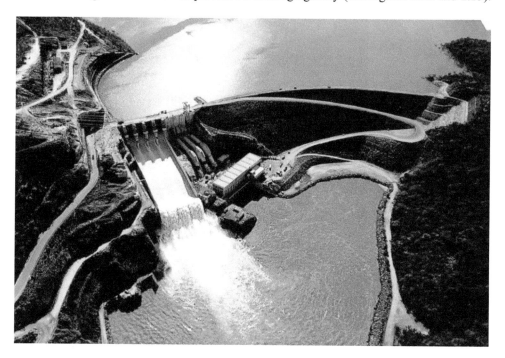

Figure 2.22 Corumbá HPP, Brazil.
Source: (CBDB, Highlights, 2006).

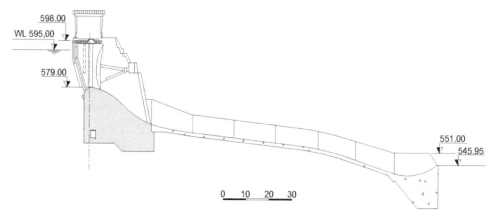

Figure 2.23 Corumbá chute spillway.
Source: (CBDB, MDB, 2000).

Serra da Mesa HPP, Tocantins river (GO, Brazil)

Spillway on the right abutment. Discharge capacity $Q = 15.000$ m³/s, with five segment gates of 15 m × 19 m. The spillway chute is over the sound granite saving concrete volume (see Figures 2.24 through 2.26).

Figure 2.24 Serra da Mesa rockfill dam. Underground powerhouse – left abutment. Spillway – right abutment.

Figure 2.25 Serra da Mesa dam. Downstream view of the spillway.

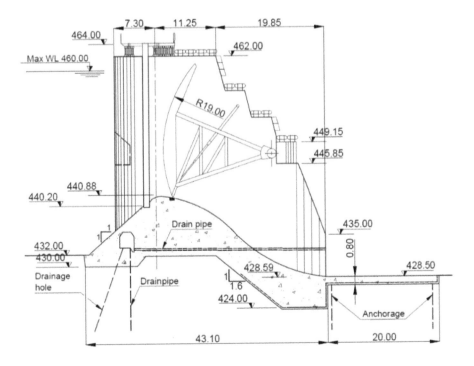

Figure 2.26 Chute spillway.

Source: (CBDB, 2000).

2.7   Culvert spillway (bottom outlet)

A culvert spillway is normally controlled by a gate and is used when a large flow capacity is required in low head reservoirs (DSD, p. 430). The bibliography cites that it is also used as a sand trap. The hydraulic design of this type of spillway is done as presented in section 3.6.1. The structure must be checked in a reduced model.

The Sobradinho plant, São Francisco river (BA, Brazil), has two spillways, one of which is a culvert spillway with 12 gates, with a hydraulic energy dissipator (Figures 2.27 to 2.29). The culvert spillway allows water to be passed to the generation in the downstream plants during the construction period (Figure 2.29B). The Sobradinho reservoir regulates downstream a minimum flow of 2.060 m^3/s, with a depletion of 12 m.

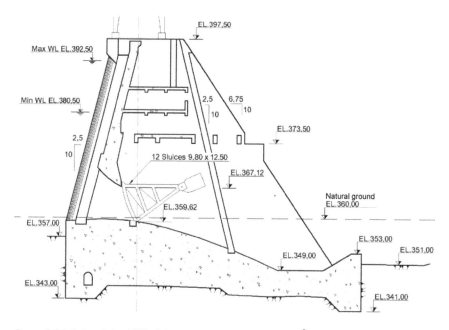

Figure 2.27 Sobradinho HPP. Culvert spillway, Q = 16.000 m^3/s.
Source: (CBDB, 1982a).

Figure 2.28 Sobradinho HPP. Substation, right bank dam, powerhouse, surface spillway, culvert spillway, left bank dam.

Figure 2.29A Surface spillway working.

Figure 2.29B Culvert spillway working.

2.8 Orifice spillway

Spillways in orifices are used in the case of arch dams in narrow valleys. The jet is launched at a distance from the dam. Their dispersion in the air ensures a reduction of erosive capacity downstream (Lencastre, 1982). The hydraulic design is given in section 3.6.2.

Among others, mention is made of the spillway of the Cahora Bassa HPP in the Zambesi river in Mozambique (see Figures 2.30 and 2.30A–B), which will be discussed in Chapter 4 (Quintela and Cruz, 1982), with two spillways:

- a bottom spillway composed of eight orifices of $6,0 \times 7,8$ m, in two groups of four, with 81 m of head on the gates, to 13.100 m^3/s;
- a surface spillway, a span of $6,0$ m $\times$ 15,5 m, to 270 m^3/s (Lemos, 1965).

Figure 2.30 Cahora Bassa HPP. Africa. Localization map. Upstream of Kariba dam. Notice Lake Malawi.

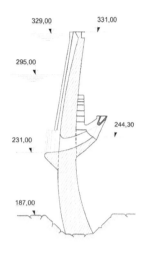

Figure 2.30A Cahora Bassa HPP (2.000 MW). Orifice
spillway/crossed jets.

Source: (Easy.Africa.com).

Figure   2.30B Cross-section.   Upper   gates
6,0 m × 15,5 m; eight bottom
segment gates 6,0 m × 7,8 m.

Source: (Spillways, Spanish Committee/1988).

Another example of this type of spillway is the Mossyrock dam on the Cowlitz river, a tributary of the right bank of the Columbia river, owned by the city of Tacoma, Washington, USA. This dam, at 185 m high, has irrigation and power generation objectives. The spillway has four gates with dimensions 15,24 m high × 12,95 m wide and a capacity of 7.800 m³/s (Figures 2.31A and B.

Figure 2.31A Mossyrock HPP.

Source: (Kollgaard and Chadwick, 1988).

To settle the dam in the rocky massif, andesite without any significant failure, the preglacial deposit of 60 m thick was removed. Thus, at the time of spillage, the jets plunge into a plunge pool at that depth. Underwater inspections conducted up until 1978 did not reveal any erosion problems.

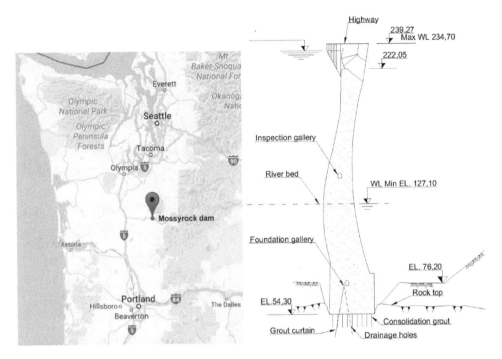

Figure 2.31B Mossyrock dam. Localization map/cross-section.

Source: Wengler (1982).

To complement the Boundary dam (1.000 MW) on the Pend Oreille river, also in Washington, USA, along the border with Canada, completed in 1967, built and operated by the city of Seattle, a solution was adopted combining this type of spillway, in orifices (seven), with surface spillways on the abutments: one on the right and two on the left (see Figures 2.32 and 2.33).

One is for debris, with a total capacity of 10.000 m³/s. The orifices have 5,1 m × 6,4 m sluices; the surface spillways have 15,2 m × 13,4 m radial gates; the spillway for debris has basculante of 7,9 m × 2,4 m. The powerhouse is underground on the left abutment and houses six Francis turbines (H = 76 m). The dam is 105 m high and 226 m long at the crest. The local geology in the Rocky Mountains is composed of Precambrian metamorphic rocks, made up of dolomitic sandstones.

2.9 Labyrinth spillway

A labyrinth spillway has a crest with broken axis (Figures 2.34 and 2.35), a solution for extending the crest length for a given fixed width available at the site and consequently increasing the spilling capacity of the structure.

These spillways are particularly advantageous in situations where it is necessary to discharge a large amount of water with a small charge on the crest. The maintenance cost is low when compared to spillways with gates. They have been designed and built for various

Downstream view (www.hydro.org).

Figure 2.32 Boundary HPP.

Source: (Kollgaard and Chadwick, 1988).

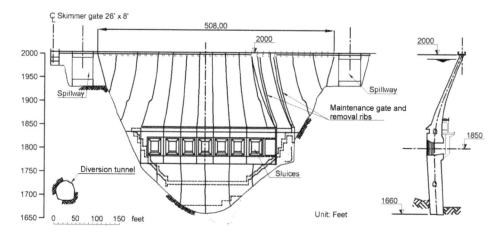

Figure 2.33 Upstream view and cross-section.

Source: (Kollgaard and Chadwick, 1988)

Figure 2.34 Bonfante HPP. Paraibuna river. Brazil.

flow capacities and are suitable for existing spillway rehabilitation projects. The structure is quite economical.

From left to right: labyrinth spillway, spillway with gates and plant building. The intake, powerhouse and the restitution tunnel are covered by the highway BR-040 on the border of the states of Rio de Janeiro and Minas Gerais. On the left bank of the reservoir you can see the railroad Belo Horizonte-Rio de Janeiro. The restitution tunnel was built next to the railroad tunnel.

The hydraulic design is presented in section 3.6.4. These spillways, in addition to increasing flow capacity for a given slope elevation, can be used to increase storage capacity while maintaining spillway capacity.

Figure 2.35 Bonfante HPP. Labyrinth spillway working.

2.10   Siphon spillway

These spillways are used when it is necessary to automatically spill an excess of inflow to keep the reservoir water level at a specified elevation. It is the ideal emergency spillway structure in remote locations. The most common standard type is shown in Figure 2.36.

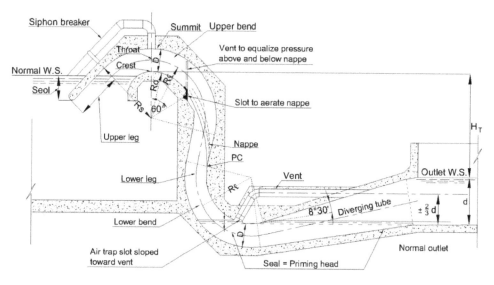

Figure 2.36 Low head siphon spillway.

Source: (USBR, Design of Small Dams, 1974).

The flow is given by $Q = CA\sqrt{2gh}$, where C = 0,9 and is illustrated in Figure 2.36 A (Linsley and Franzini, 1955).

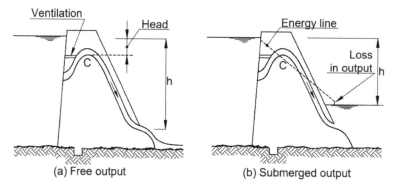

Figure 2.36A Siphon spillway.

Source: (Linsley and Franzini, 1955).

Figure 2.37 A shows the siphon spillway of the Ohau Lake dam, New Zealand, designed to keep a minimal outflow on the Ohau river.

Figure 2.37A Siphon spillway, Ohau river dam.

Source: (Antony Burton, New Zealand).

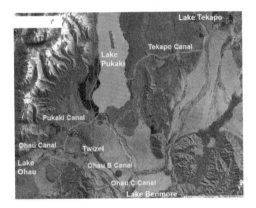

Figure 2.37B The map of Ohau/Pukaki canal. Lake Ruataniwha downstream.

Source: (Fishingmag.co, New Zealand).

It is opportune to show (Figure 2.37C) the hydraulic works of the Ohau and Pukaki channels, adding water to the Ohau B HPP, which flows into lake Ruataniwha near the town of Twizel.

2.11   Stepped spillway

With the introduction of roll compacted (CCR) dams, it has become convenient to leave steps on the downstream side of the cross-section of the spillway of a gravity dam.

The steps dissipate energy and reduce the excess to be dissipated in the basin, bringing substantial economic benefits to the project, as it shortens the length of the dissipation basin.

The stepped spillways design procedure is presented in section 3.6.3.

Figures 2.38 and 2.39 shows the solution adopted in the Dona Francisca HPP (125 MW), developed by Engevix Engeneering for Dona Francisca Energética S.A in the Jacuí river (Brazil, state of Rio Grande do Sul).

The spillway, designed to 12.340 m³/s with a head of 6 m, was studied in detail in a reduced hydraulic model in CEHPAR (Curitiba, Paraná state).

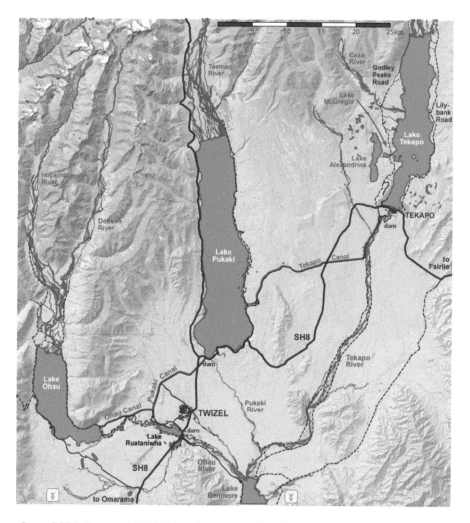

Figure 2.37C Ohau and Pukaki lakes flowing into lake Ruataniwha on the Ohau river, near the town of Twizel. New Zealand (Localization map).

Source: Antony Burton.

It is worth mentioning the La Grande 2 spillway (LG-2) on the La Grande river that flows into James Bay. Figures 2.40 A and 2.40 B show all the plants of the La Grande Complex, 16.021 MW, built between 1970 and 1990 by owner Hydro Québec.

The rocky massif on the site consists of a very hard granite of the Canadian Shield. The main dam is of rockfill with a classic clay waterproof core. It is 162 m high and 2.835 m long. According to Hayeur Gaëtan (2001), the implementation of the dam caused the following environmental impacts: it altered the course of 17 rivers and required the creation of 28 lakes and six basin transpositions (Caniapiscau, Laforge, Eastmain, Boyd-Sakami, Opinaca and Rupert).

Figure 2.38 Dona Francisca HPP. Stepped spillway, 355 m length, and stilling basin.
Source: (CBDB, Highligts, 2006).

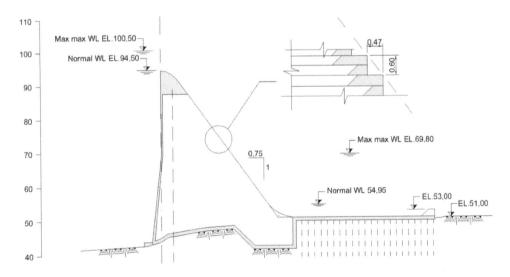

Figure 2.39 Dona Francisca HPP. Stepped spillway and stilling basin.
Source: (CBDB, MDB, 2000).

The spillway, with a maximum capacity of 17.600 m³/s, has 13 steps of 10 m high to dissipate the energy along a channel of 1,5 km in length (Aubin *et al.*, 1979). The solution in steps on the rock is very economical.

A – Localization map.

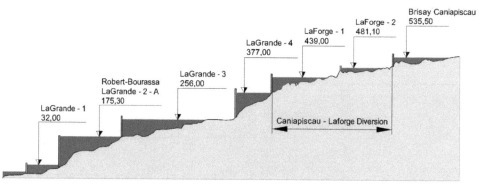

B – Inventory Studies – River Profile, Fall Division.

Figure 2.40 La Grande Complex 2, La Grande river, Québec, Canada.

C – La Grande 2 HPP. The rockfill dam and the spillway.

D – Stepped spillway working.

E – Detail of energy dissipation on the steps.

Figure 2.40 (Continued)

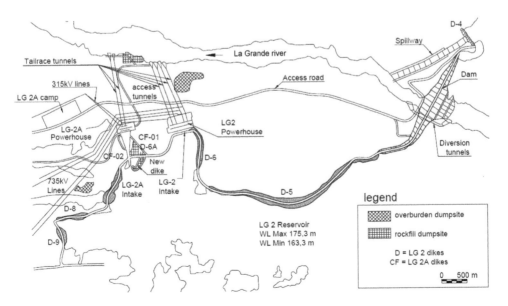

Figure 2.41 General arrangement of the works of the La Grande 1, 2 and 2 A.
Source: (WP&DC, 01/1990).

2.12   Horizontal apron spillway (spillway without ogee)

Here is presented the case of the Baguari HPP spillway, a horizontal apron without ogee ($P = 0$). This kind of spillway has a low flow coefficient, of the order of 1,7 (see Figure 3.10). This plant (Figures 2.42 to 2.45) is located in river Doce, close to the city of Governador Valadares, state of Minas Gerais, Brazil. The spillway has six gates of 17,0 m × 19,1 m; $P = 0$, $H = 18,6$ m and a flow capacity of 13.000 m^3/s. It was verified in hydraulic model, resulting in a coefficient of flow $C = 1,6$.

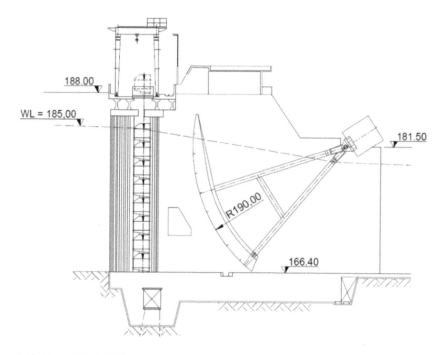

Figure 2.42 Baguari HPP Spillway.

Figure 2.43 River diverted by spillway.   Figure 2.44 View of the Baguari Plant.

Figure 2.45 Baguari Plant. Aerial view. Observe the tranquility of the flow approaching the structures of the spillway and intake near the left bank.

Chapter 3

Spillway design

The hydraulic design of the spillway basically consists of determining the width (L) and height of the hydraulic head on the crest (Ho) to safely discharge the design flood. The stilling basin for energy dissipator design will be presented in Chapter 5.

Figure 3.1 UHE Tucuruí. Tocantins river. State of Pará. Brazil.
Source: Owner: Eletronorte.

In this book the case of the Tucuruí spillway (Figure 3.1) will be explored, since the author has much data about the project, having worked on it for five years (from 1981 to 1986), including the elaboration of the operation plan for the gates and the training of the operation team.

3.1 Initial considerations

The hydraulic design of spillways has been done using the guidelines and the design criteria of the USBR for more than 60 years. Normally, the design must be supported by studies in hydraulic models, one three-dimensional and another two-dimensional.

The tests in the three-dimensional model are fundamental for the verification and optimization of the efficiency of the structure with respect to the layout of works, considering the joint operation with the adjacent structures of the intake and powerhouse.

It is essential that the spillway receives the inflow as quietly as possible, without instabilities and detachments or disturbances/depressions of the liquid vein, which may lead to unsatisfactory performance of the structure, reducing its discharge capacity, or implying pressure fluctuations that may result in a cavitation tendency in the structure, as will be discussed in more detail in Chapters 4 and 8 (see Figure 3.1D).

Chapter 4 deals with the question of the flownet and the hydrodynamic pressures over the spillway. The tests allow verification of whether the flow will be restored to the riverbed in suitable form, with security, minimizing the possibilities of scour downstream. The tests in the bidimensional model (Figure 3.1E) make it possible to verify and measure the hydrodynamics pressures on the structure, as well as to test and to detail the terminal structure of energy dissipation.

Chapters 5, 6 and 7 deal with the energy dissipation, with the efforts downstream of dissipators and with the evaluation of scour downstream.

Experience shows conclusively that hydraulic studies on reduced models provide substantial savings for the power plant project that pay off model studies. They make it possible to investigate and optimize the lengths and heights of side walls, as well as the length of the bottom slab and the terminal dissipation structure, often resulting in significant reductions in concrete volumes and consequently in the cost of the plant.

3.2   Ogee shape

The initial crest forms were usually based on a simple parabola (Figure 3.1A) designed to fit the trajectory of a projectile. According to this principle, it is assumed that the horizontal velocity component of the flow is constant and that the only force acting on the nappe is gravity. For the derivation of nappe profiles over sharp-crested weir by the principle of projectile, see Chow (1959).

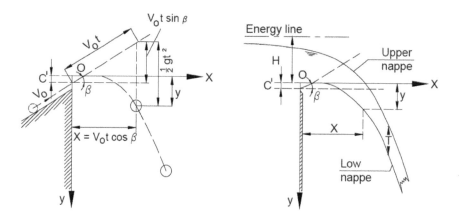

Figure 3.1A Nappe profiles over sharp-crested weir by the principle of projectile.

Bazin made the first laboratory tests to investigate nappe shapes from 1886 to 1888. The use of this author's data in the design produced a crest shape that coincides with the bottom surface of an aerated nappe on a weir plate.

The adoption of such profile, generally known as a Bazin profile, theoretically should cause negative crest pressures (as shown in Figure 3.1 B), which can cause the spill to detach from contact with the concrete surface of the spillway. In selecting the profile, these negative pressures should be avoided.

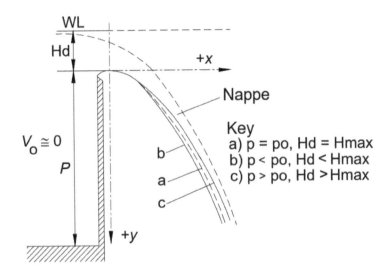

Figure 3.1B Variations in nappe shape.

The profile is designed for the design head (H_d). For this head the pressures will be close to the atmospheric ($p = p_o$) and the relative pressures will be zero. For higher heads than the design head, negative pressures (lower than atmospheric pressure) will occur. For lower heads than the design head, positive pressures will occur (above atmospheric pressure).

According to Chow (1959), the profile of Bazin was modified by several researchers: Bazin (1898), Marchi (1928), Scimemi (1937), Escande (1929), Creager (1945), USBR (1948) and Smetana (1949).

According to Rouse (1938), Bazin (1898) and Rehbock (1929) proposed formulas for rectangular spillway without lateral contraction.

The shape of the nappe was established by Scimemi (1937), measured from a source at the highest point of the crest, for a unit value of H, as a function $Y = K x^n$, where $K = 0.5$ and $n = 1,85$ (section 3.3). The nappes for other H values are similar, and this function has been rewritten.

From 1932 to 1948. the USBR conducted extensive trials on the shape of the nappe over a spillway. For practical purposes, the US Army Corps of Engineers has developed several standard forms in Waterways Experiment Station, which are summarized in the HDC (1955/1959).

Spillways differ from weir plates in two important respects. From a dynamic point of view, the lower surface of the nappe is in contact with a fixed limit, along which the pressure intensity is neither necessarily constant nor atmospheric. From an economic point of view, a spillway with a given length must safely discharge a peak of flow under the smallest possible head.

The American engineers standardized the crest of the spillway based on the shape of the weir plate in order to obtain maximum flow without reducing the pressure below atmospheric pressure.

Tests performed by Rouse (1938) showed that the spillway designed in this way will produce a nappe that is not affected by the presence of a fixed lower boundary, provided the spillway shape has the same shape of the nappe for the same discharge. As the head varies, the pressures will also vary, as shown in Figure 3.1 C. If the spillway profile is smoothly curved, according to the nappe shown, the head can be increased without danger of nappe separation. This applies only to the case of two-dimensional flow.

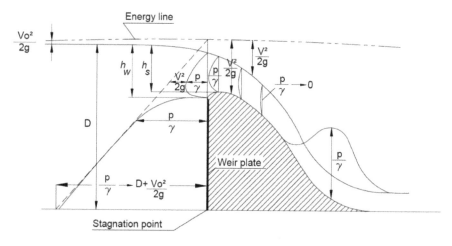

Figure 3.1C Pressure distribution in a spillway designed on the basis of the ventilated nappe of a weir plate.

Source: (Rouse, 1938).

However, Escande (1929) has shown that abrupt curvatures at the crest can cause separation of the flow with consequent pressure drop, which is always undesirable. This subject was detailed by Rouse (1938). The Figure 3.1D show aspects of separation.

Figure 3.1E shows the reduced two-dimensional model for detailing the Tucuruí HPP spillway. Observe that the flow lines are accommodated, adhering to the surface of the concrete, without discontinuity or separation, for the flow of 100,000 m³/s (Hidroesb, Final Report, 1982).

It is clear the importance of carrying out the model tests for the careful verification of these aspects, among others.

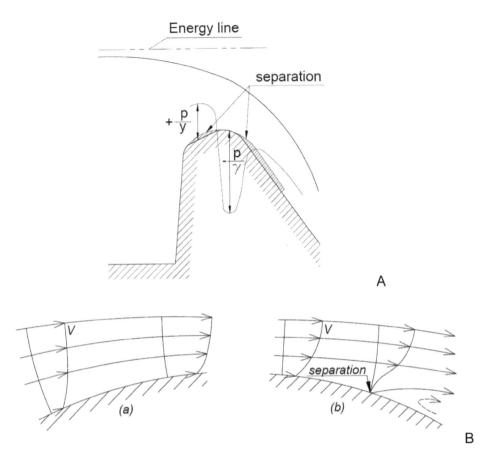

Figure 3.1D A – Separation of the flow at the crest of a poorly designed spillway. B – (a) acceleration effect, (b) deceleration effect on a velocity distribution.

Source: (Rouse, 1938).

Figure 3.1E Spillway of Tucuruí HPP. Two-dimensional reduced model. Flownet for Q = 100.000 m³/s, without separation.

Source: (Hidroesb, Final Report, 1982).

3.3 Hydraulic design

As previously stated, the hydraulic design of the spillway consists in determining the necessary dimensions of this structure to safely flow the design discharge, that is: the width (L); and the hydraulic head on the crest (Ho); It is observed the design head is often called H_d.

The hydraulic design of the uncontrolled still considers a standard crest – Figure 3.2.

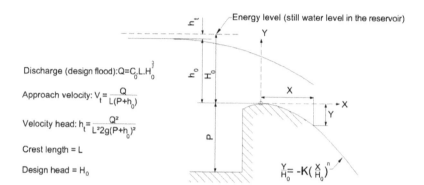

Discharge (design flood): $Q = C_0.L.H_0^{\frac{3}{2}}$

Approach velocity: $V_t = \dfrac{Q}{L(P+h_0)}$

Velocity head: $h_t = \dfrac{Q^2}{L^2 2g(P+h_0)^2}$

Crest length = L

Design head = H_0

Figure 3.2 Standard ogee crest – sketch.
Source: Chow (1959).

This profile depends on the design head Ho, the inclination of the upstream face and the crest height P of the entrance channel. These parameters influence the velocity of the approach flow Vt, as well as the flow coefficient C.

The equation of the downstream face is:

$$\frac{Y}{Ho} = -K\left(\frac{X}{Ho}\right)^n \tag{3.1}$$

where

K and n are constants that depend on the inclination of the upstream face and of the approach velocity. Figures 3.3 and 3.4 (Chow, 1959) provide the values of these constants for different slope conditions of the upstream slope.

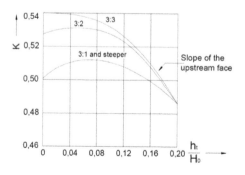

Figure 3.3 Values of k, function of the approach velocity head.

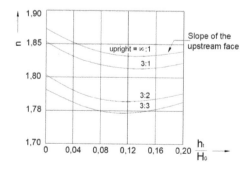

Figure 3.4 Values of n, function of slope of upstream slope and velocity of approach.

The USBR – HDC 111–1 and 111–2/1 charts (HDC, 1959), obtained in extensive testing campaigns, present the geometries of the upstream and downstream quadrants of this standard profile in a standardized way (see Figure 3.5).

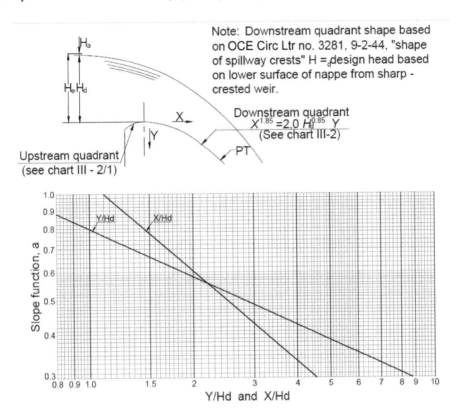

Figure 3.5 Overflow spillway crest, tangent coordinates.

Source: Hydraulic Design Chart 111–1 (HDC, 1959).

The determination of *P* and *Ho* is done as a function of the boundary conditions of each project. In practice, each project is different, because the conditions vary from place to place: the elevations of the reservoir and of the foundation of the spillway vary.

In the Basic Project, one has to use the elevation of the reservoir fixed in the Inventory Studies of the river that, according to the legislation in Brazil, cannot be modified.

In the example of Tucuruí HPP, presented in section 3.5, it was:

- the bottom of the Tocantins river on the site was El. 0,00 m;
- the normal WL of the reservoir was El. 72,00 m (as will be shown later, the design flood may vary from 90.000 to 110.000 m³/s).

That is, *P* + *Ho* = 72 m. It should be noted that the *P/Ho* ratio directly influences the spillway flow coefficient, as will also be shown in Figure 3.10.

In the final solution designed for the spillway design, the crest of the spillway was set at elevation 52.00 m. So, $P = 52$ m and, consequently, $Ho = 20$ m. This would be mean gates at least 20 m high, at that time very high (see section 3.5).

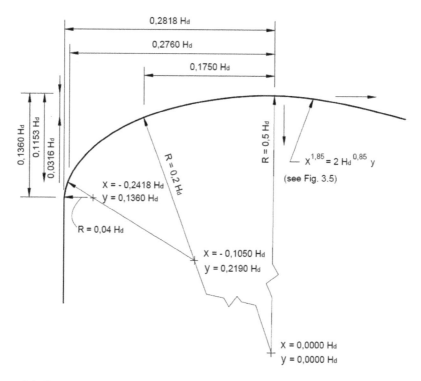

Figure 3.6 Crest geometry, upstream quadrant.
Source: Chart HDC 111–2/1 (1959).

The discharge capacity of an uncontrolled spillway is given by the equation:

$$Q = C_o L H_o^{3/2} \tag{3.2}$$

where:
Q = total discharge (m³/s);
C_o = discharge coefficient;
L = effective crest length (m);
H_0 = hydraulic head (m).

It should be emphasized that the higher the discharge coefficient, the smaller the spill-way width, which makes it possible to save volume of concrete – the most expensive civil works service. Every effort should be made to optimize the discharge coefficient.

The design head H_o is the head used to define the downstream profile of the spillway.

The hydraulic head (H_d) may be equal to, less than or greater than H_o. For $H_d > H_o$, the discharge coefficient increases and negative pressures occur at the face of the spillway.

In determining the net length of the crest (L), the contraction effect of the piers K_p and the extreme walls K_a (abutments) must be taken into account, as defined in Charts 111–3/1 and 111–3/2 of the HDC (1959) shown in Figures 3.7 to 3.9.

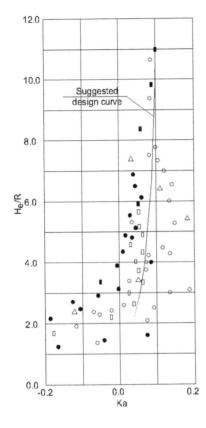

Basic equation

$$Q=C[L'-2(NKp+Ka)He]He^{3/2}$$

Where:

Q = discharge, cfs
C = discharge coefficient
L' = net length of crest, ft
N = number of piers
Kp= pier contraction coefficient
Ka= abutment contraction coefficient
He= total head on crest, ft

Legend

Symbol	Project	R	W/L	W/H
○	CW 801	4	1.55	0.96
●	FOLSOM	8	2.10	3.77
◻	PHILPOTT	5	2.67	1.42
◼	PINE FLAT*	4	2.12	1.77
△	CENTER HILL*	5	3.83	9.48

*Gated spillway with piers

Note:

R = radius of abutment, ft
W = width of approach reproduced in model, ft
L = gross width of spillway, ft
H = depth of approach in model, ft

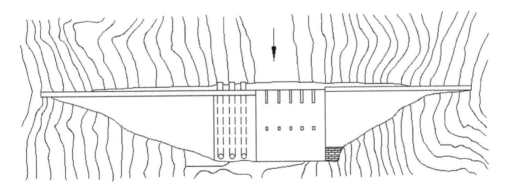

Figure 3.7 Overflow spillway crest with adjacent concrete structures. Abutment contraction coefficient (Ka).

Source: (HDC 111–3/1, 1959).

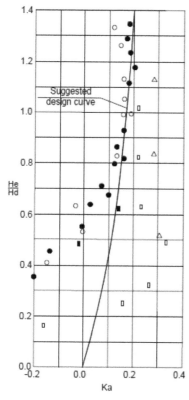

Basic equation

$$Q = C\left[L' - 2(NK_p + K_a)H_e\right]H_e^{3/2}$$

Where:

Q= discharge, cfs
C= discharge coefficient
L'= net length of crest, ft
N= number of piers
Kp= pier contraction coefficient
Ka= abutment contraction coefficient
He= energy head on crest, ft

Legend

Symbol	Project	R	W/L	W/H
○	DORENA	2	5.60	10.7
●	DORENA	4	5.60	10.7
□	RED ROCK*	7.8	3.42	16.5
■	CARLYLE*	9	8.44	75.5
△	WALTER F. GEORGE*	4	5.44	55.3

*Gated spillway with piers

Note:
R = radius of abutment. ft
W = width of approach reproduced in model, ft
L = gross width of spillway, ft
H = depth of approach in model, ft
H_d = design head on crest, ft

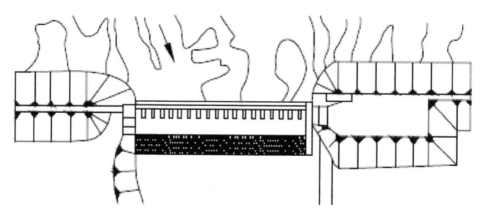

Figure 3.8 Overflow spillway crest with adjacent embankment sections. Abutment contraction coefficient (Ka).

Source: (HDC 111–3/2, 1959).

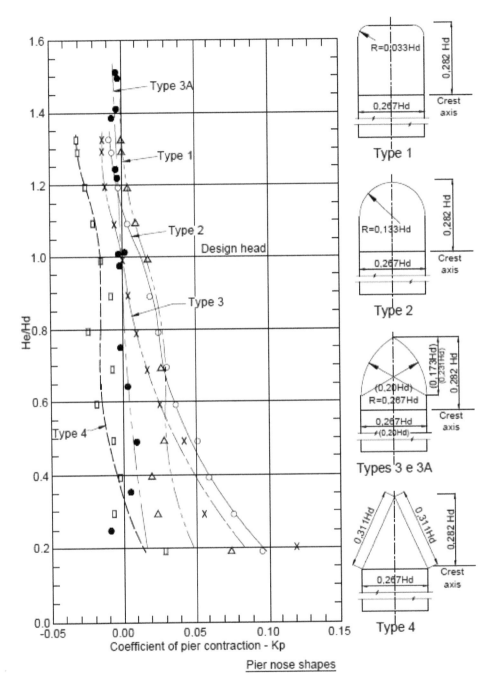

Figure 3.9 Gated overflow spillways. Pier contraction coefficients. Effect of nose shape.
Source: (HDC 111–5, 1959).

$$L' = L + 2(NK_p + K_a)H_o \tag{3.3}$$

where:
L= effective crest length (m);
L'= net crest length (m);
N = number of piers;
K_a = coefficient of contraction of the extreme walls of the spillway (Figures 3.7 and 3.8).
K_p = coefficient of contraction of the piers (Figure 3.9);
The discharge coefficient is influenced by a number of factors, such as:

*   approach flow;
*   depth of the inlet channel – P (Figure 3.10);
*   loads different from the design head (Figure 3.11);
*   slope of the upstream facing (Figure 3.12);
*   effects of downstream submergence (Figure 3.13).

With respect to the tranquility of the approach flow, the ideal arrangement of the works is one in which the structures are positioned on an axis normal to the direction of the main flow of the river. However, this is not always possible. When reduced models are available for experimental studies, one can optimize the position in the plan of the structures in order to have a smooth approach flow to the spillway, without detachments and disturbances, in such a way that the nappe is as stable as possible. In the three-dimensional model, the interferences of the functioning of the adjacent structures in the behavior of the flow in the extreme spans of the spillway are verified and optimized. The determination of the flow coefficient is facilitated.

When hydraulic models are not available for experimental studies, in the launching of the structures of the arrangement one must, based on experience, try to draw a flow network considering the topographical conditions of the contour of the place. In this regard, a key reference is Dao-Yang and Man-Ling (1979), *Mathematical Model of Flow Over Spillway Dam.*

In section 10.5 are presented the studies in the model of the spillway of the Tucuruí HPP, with a design discharge of 110.000 m³/s. This structure is positioned between the intake and the right bank dam (Figures 3.16, 3.17 and 3.18A).

It should be noted once again that the savings obtained for the projects in the studies on hydraulic models in terms of reduction of concrete volume of side walls of spillways and of stilling basins' lengths are substantial and pay the cost of these models.

In determining the discharge coefficient of the spillways controlled by gates, Figure 3.14 and the DSD (1973), Chapter IX, Chart 200, should be noted.

For side channel spillways, note what is stated in Chart 202 – DSD (1973).

For shaft spillways, also called morning glory spillways, one should observe Chart 212 – DSD, and also the one shown in Charts 140–1 to 140–1/8 of HDC (1959).

In Figure 3.10, it should be noted that the flow coefficient varies markedly for the range of values $0 < P/Ho < 1, 0$. For $P = 1.0$, it has $C = 2,15$. For $P = 0$, $C = 1,7$. $P = 0$ means that the spillway has no sill (it is a slab). Examples:

*   Case of the Baguari HPP (Figure 2.42), in the Doce river, slab spillway ($P = 0$), six gates of $17,0 \times 19,1$ m for 12.800 m³/s. In the model, the flow coefficient was estimated to be 1,6.

• Case of the Dona Francisca HPP (Figure 2.39), uncontrolled spillway (without gates), Sobrinho (2000) presented the values shown in Table 3.1.

Table 3.1 Flow coefficients of the spillway of HPP Dona Francisca

Q m³/s	L (m)	q (m³/s/m)	H (m)	C
12.340	355	36,84	5,97	2,38
8.500	355	23,94	4,80	2,28
6.080	355	17,13	3,93	2,20
4.500	355	12,68	3,27	2,14
1.500	355	4,22	1,68	1,94

• Case of the Tucuruí HPP (section 2.4):

Table 3.2 Flow coefficients of the spillway of HPP Tucuruí

Q m³/s	L (m)	q (m³/s/m)	H (m)	C
110.000	460	239,13	23,3	2,13
100.000	460	217,39	22,0	2,11
86.500	460	188,04	20,0	2,10

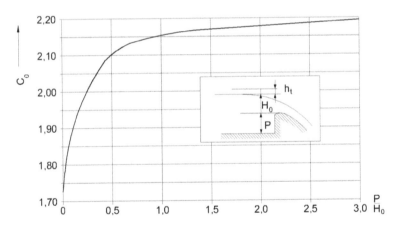

Figure 3.10 Coefficients of discharge × relation P/Ho.
Source: Chow (1959).

The waterline profiles on the spillway can be determined using the Charts III-11 to III-15 of HDC (1959), Figure 3.15 and Table 3.3.

3.4 Practical rules for defining the type of spillway

The first stage of studies of a hydrographic basin, in Brazil the Hydroelectric Inventory Studies, defines how many plants may be built in each river and the water levels of each plant, as well as the preliminary arrangements of each one.

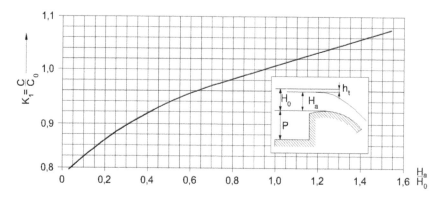

Figure 3.11 Correction of discharge coefficient for other heads beyond the design head (*Ho*).
Source: Chow (1959).

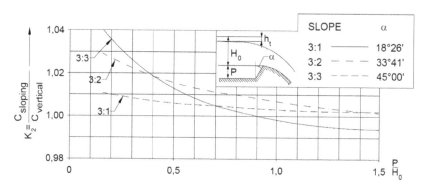

Figure 3.12 Coefficient of discharge for ogee with inclined upstream facing.
Source: Chow (1959), (Lysne, 2003).

In the next stage of Feasibility Studies, in principle, water levels cannot be changed. Any future changes in water levels that are extraordinarily necessary, can only be made if approved by the National Electric Energy Agency, ANEEL, in Brazil.

As stated in section 1.2.1, the topographical and geological characteristics of the site condition the types of arrangements, which may be compact or of derivation.

In compact arrangements of the plants, the dam structures are usually positioned/aligned on a single axis.

In derivation arrangements, the structures of the hydraulic circuit of power generation, intake and powerhouse are positioned on another axis, usually on one of the abutments.

3.4.1   *Location of the spillway*

A   *Plants located in wide plains*

In plants located on wide plains, usually the spillway is positioned in the axis of the dam's structures, as shown in the example of the Tucuruí HPP.

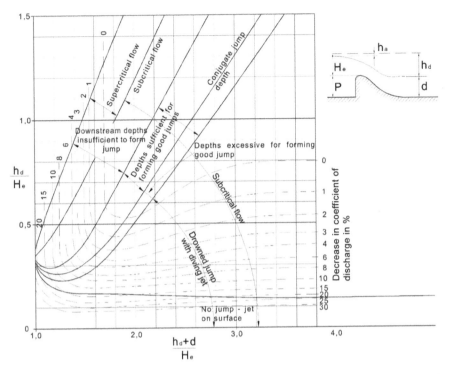

Figure 3.13 Correction of discharge coefficient in function of the downstream effects (basin elevation and submergence).

Source: (DSD, 1974, illustration 252, p. 380). (Schreiber, 1978, Diagram 6, p. 82).

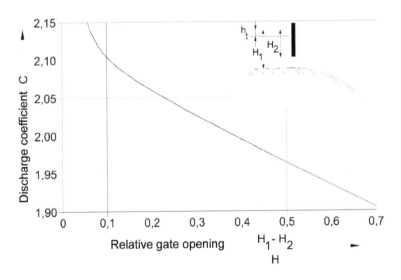

Figure 3.14 Coefficients of discharge for controlled flow. Obs.: See note on the position of the gate – Chapter 9.

Source: Chow (1959).

Table 3.3 Coordinates for the upper nappe

-	No piers			Center of the bay between piers (type 2)			Along piers		
H/H_d	0,50	1,00	1,33	0,50	1,00	1,33	0,50	1,00	1,33
$\frac{X}{H_d}$	$\frac{Y}{H_d}$	$\frac{Y}{H_d}$	$\frac{Y}{H_d}$	$\frac{Y}{H_d}$	$\frac{Y}{H_d}$	$\frac{Y}{H_d}$	$\frac{Y}{H_d}$	$\frac{Y}{H_d}$	$\frac{Y}{H_d}$
−1,0	−0,490	−0,933	−1,210	−0,482	−0,941	−1,230	−0,495	−0,950	−1,235
−0,8	−0,484	−0,915	−1,105	−0,480	−0,932	−1,215	−0,492	−0,940	−1,221
−0,6	−0,475	−0,893	−1,151	−0,472	−0,913	−1,194	−0,490	−0,929	−1,209
−0,4	−0,460	−0,865	−1,110	−0,457	−0,890	−1,165	−0,482	−0,930	−1,218
−0,2	−0,425	−0,821	−1,060	−0,431	−0,855	−1,122	−0,440	−0,925	−1,244
0,0	−0,371	−0,755	−1,000	−0,384	−0,805	−1,071	−0,383	−0,779	−1,103
0,2	−0,300	−0,681	−0,919	−0,313	−0,735	−1,015	−0,266	−0,651	−0,950
0,4	−0,200	−0,586	−0,821	−0,220	−0,539	−0,944	−0,185	−0,545	−0,821
0,6	−0,075	−0,465	−0,705	−0,088	−0,647	−0,847	−0,076	−0,425	−0,689
0,8	0,075	−0,320	−0,569	−0,075	−0,389	−0,725	0,060	−0,285	−0,549
1,0	0,258	−0,145	−0,411	−0,257	−0,202	−0,564	0,240	−0,121	−0,389
1,2	0,470	0,055	−0,220	0,462	0,015	−0,356	0,445	0,067	−0,215
1,4	0,705	0,294	−0,002	0,705	0,266	−0,102	0,675	0,286	−0,011
1,6	0,972	0,563	0,243	0,977	0,521	0,172	0,925	0,521	0,208
1,8	1,269	0,857	0,531	1,278	0,860	0,465	1,177	0,779	0,438

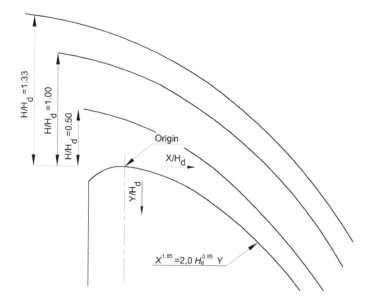

Figure 3.15 Waterline profiles. (HDC III-11, III-12, III-13). Without piers (HDC III-11)

The Sobradinho plant, on the São Francisco river in the northeastern region of Brazil, has two spillways: one on the surface and the other on the bottom. The bottom (culvert) spillway was inserted to allow the passage of water during the construction period to enable power generation in the existing plants downstream.

B *Plants located in stretches of moderately embedded rivers*

In a plant situated in a moderately enclosed valley, the spillway can be positioned on the abutment: examples – Itaipu HPP, Corumbá HPP and Serra da Mesa HPP, among others. In the latter, a function of good geology, the spillway is a low and short sill at the right abutment, followed by an unlined chute (Figures 2.25 and 2.26).

Other spillways, such as Xingó HPP, Machadinho HPP (see section 5.4) and Nova Ponte HPP, also have part of the chute unlined, bringing substantial savings of concrete to the project.

C *Plants located in very embedded valley – "V" valley*

In the case of a plant situated in a "V" valley, the spillway can be positioned:

- over the dam, like Karakaya HPP, Mossyrock HPP and Kariba HPP, among others;
- on the abutment, like Irapé HPP spillway in a tunnel, Hoover HPP in two side spillways/tunnels, and Monjolinho HPP in a lateral spillway. In this type of place, one can have also a morning glory spillway, like Colino 1 SHP.

3.4.2 *Spillway main parameters: P, Ho, P/Ho and Co*

Having chosen the location of the plant, the foundation elevation, the water level of the reservoir elevation and the discharge of the spillway project, one has the elements to do the project. As previously mentioned, P = height of the sill, Ho = design head of the structure (and of the gate).

It should be noted that the P/Ho ratio greatly influences the coefficient of discharge of the spillway (Figure 3.10) and that the flow capacity varies with $Ho^{3/2}$. Therefore, these aspects and the shapes of the abutments condition the hydraulic design.

The boundary conditions of the site are very important. In a narrow valley, one usually has to increase Ho to reduce the width of the spillway.

By the way, it is recorded that gate width (L) is usually smaller than Ho. In May 2014, the author consulted Erbisti (author of the book *Design of Hydraulic Gates*) on the relation Ho/L = 1,2. According to him, it is not a rule but a recommendation to have the following advantages:

1 since the main beams have a horizontal direction, the smaller the span of support of the beam, the smaller its weight, as well as the weight of the gate;
2 gates that are higher than they are wide are easier to move inside vertical guides; thus, the probability of being stuck or torsion during its movement decreases.

Each case is different and has its peculiarities. The author recommends that the engineers should use existing experience.

The following are notes on the design of the Tucuruí HPP spillway, which has $P = 52$ m and $Ho = 22$ m and so the coefficient of discharge $Co = 2.18$. In Tucuruí, this coefficient was checked in two hydraulic models.

Reminder: it should not be forgotten that if $P = 0$ the coefficient of discharge would fall to 1.7.

3.5   Hydraulic design of the ogee spillway – example of Tucuruí

This section presents some notes about the spillway of Tucuruí HPP, one of the largest in the world, in which the author had the honor to work for five years (1981 to 1986), in the hydraulics division with many interfaces with geotechnics. His participation was extended until he completed the training of the operation team of the 23 spillway gates at the plant (from 1984 to 1986).

Downstream view to the right bank.

Downstream view to the left bank.

Figure 3.16 Spillway of the Tucuruí HPP. After the spillway, note the Powerhouses I and II and the dam on the left bank.

Source: (Courtesy of Eletronorte).

3.5.1   *Design flood*

The evolution of the statistical studies of floods for the project of Tucuruí HPP project is presented in Table 3.4.

Studies have been developed for the definition of probable maximum precipitation (PMP), which leads to the determination of maximum probable flood. In these studies, the methodology recommended by the World Meteorological Organization (WMO) was used and also was used the model SSARR – streamflow synthesis and reservoir regulation.

The maximum probable flood was defined as 105.800 m^3/s. However, due to the complexity of the alternatives of meteorological possibilities involved, the flow rate of 110.000 m^3/s

Table 3.4 Floods – Gumbel (Technical Memory, 1989)

TR (years)	Discharges (m³/s)		
25	45.800	48.100	52.400
50	50.900	53.500	58.600
100	56.000	58.900	64.700
200	61.100	64.200	70.800
500	67.700	71.200	78.900
1.000	72.800	76.600	85.000
10.000	89.600	94.300	105.300

was adopted as the ultimate capacity. As recorded in the Technical Memory of the project, the hydraulic design of the spillway was made for discharge $Q = 100.000$ m³/s with a design head $Ho = 22$ m.

3.5.2   Discharge coefficient and the number of gates

In the Inventory Studies, the normal maximum water level of the reservoir was fixed at El. 74,00 m, which implies the first alternative of elevation of the crest of the spillway: 52,00 m.

Due to the elevation of the riverbed in the region upstream of the spillway, El. 0.00 m, the upstream depth of water $P = 52,00$ m was characterized in relation to the crest of the spillway.

Therefore, $P/Ho = 2,36$. Entering this value in the graph of Figure 3.10, we have the coefficient of discharge $Co = 2,18$.

Using the expression (3.1) we have: $100.000 = 2,18 \times L \times 22^{1,5}$, which results in a required total width $L = 445$ m. Due to safety, a total width of 460 m was adopted in the project. Gates 20 m in width were adopted (Figure 3.17), for a total of 23 gates. The final flow coefficient was 2,11 (which was checked in the model). The typical cut is shown in Figure 3.18.

After the flood of March 1980, 68.400 m³/s, flood studies were redone, and the project discharge passed to $Q = 110.000$ m³/s (ultimate capacity).

The design of the spillway was maintained. In the case of such an event, the reservoir water level must reach a maximum of 75,30 m – WL absolute maximum.

The dimensions of the gates were fixed at 21 m high × 20 m wide. It should be noted that:

• the magnitude of the project discharge implied having to design a large spillway with, of course, large gates;
• at the time, gates with heights of the order of 20–22 m were, and still are, considered high;
• after the study of alternatives, a $Ho = 22$ m was used and the width L of the spillway was calculated; for a smaller H, the width L would be larger, and consequently, the volume of concrete would be larger and the structure would be more expensive.

The study of alternative flood gates ($L \times H$) should define the most economic spillway, which results in a smaller volume of concrete. Regarding the selection of the type of gates, it is recommended to consult, among others, the book *Hydraulic Gates* by Erbisti (2002).

It is recorded that for the normal WL of the reservoir, elevation 72.00 m, with $H_d = 20$ m, the discharge capacity is of the order of 89.700 m³/s. The discharge capacity curves of the spillway are shown in Figure 3.19.

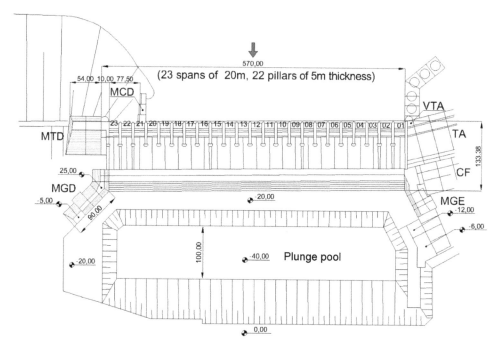

Figure 3.17 Tucuruí HPP. Spillway plant.

Source: (Magela, 2015a).

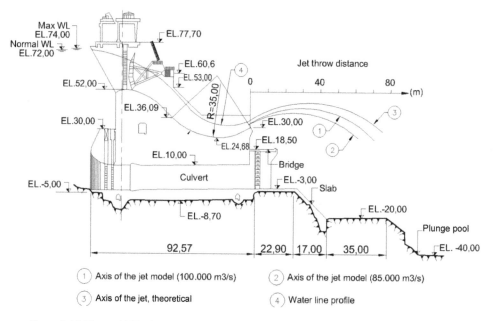

Figure 3.18 Tucuruí HPP. Ski jump spillway; 23 radial gates, 20,00 m × 20,75 m. Note trajectory length. $Q = 110.000$ m³/s → Max $WL = 75,30$ m. $Q = 100.000$ m³/s → Normal $WL = 74,00$ m. Elevation of bucket lip = 30,00 m; takeoff angle = 32°.

Figure 3.18A Tucuruí HPP. Ski jump spillway. Earth dam in the back.

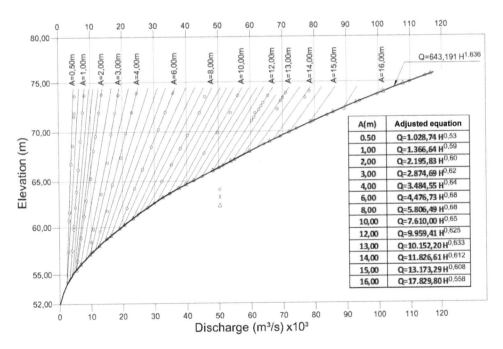

A(m)	Adjusted equation
0,50	$Q = 1.028,74\,H^{0,53}$
1,00	$Q = 1.366,64\,H^{0,59}$
2,00	$Q = 2.195,83\,H^{0,60}$
3,00	$Q = 2.874,69\,H^{0,62}$
4,00	$Q = 3.484,55\,H^{0,64}$
6,00	$Q = 4.476,73\,H^{0,68}$
8,00	$Q = 5.806,49\,H^{0,68}$
10,00	$Q = 7.610,00\,H^{0,65}$
12,00	$Q = 9.959,41\,H^{0,625}$
13,00	$Q = 10.152,20\,H^{0,633}$
14,00	$Q = 11.826,61\,H^{0,612}$
15,00	$Q = 13.173,29\,H^{0,608}$
16,00	$Q = 17.829,80\,H^{0,558}$

Figure 3.19 Tucuruí HPP. Spillway discharge curves.

Source: (Magela, 2015a).

3.5.3 Culverts for the river diversion

It should be noted that the spillway has 40 culverts for the third diversion phase, (Figures 3.18 and 3.19B). The hydraulic design was made for $Q = 56.000$ m³/s. The flow capacity was verified in two- and three-dimensional models and in the prototype. The flow coefficients calculated using Equation 3.3 are presented in the following table.

$$C = \frac{Q}{A\sqrt{2gH}} \tag{3.4}$$

where:

Q design discharge (m³/s);

A culverts area: 6,50 m × 13,0 m = 84,50 m²;

H difference between the WL upstream of the culverts and the downstream WL, considered in the elevation 10.00 m (constant) equal to the ceiling elevation of the culverts.

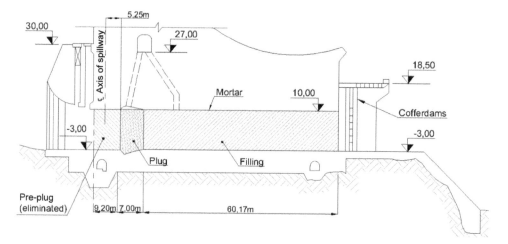

Figure 3.19B Tucuruí HPP. Spillway cross-section. Culverts.
Source: (Carvalho, 1986).

From the data of the graph (Figure 3.21), the water levels in the prototype upstream of the culverts were calculated for discharges larger than 42.000 m³/s (Frame 3.2).

It is recorded that the levels obtained in the model for these flows were approximately 2.0 meters higher, which impacted the cofferdams elevations of the third phase of river diversion, shown in the following illustrations (Figures 3.22 and 3.23).

Figure 3.23 shows the construction of the closing pre-cofferdam of the channel on the right bank to divert the Tocantins river through the 40 culverts (Figure 3.22).

3.5.4 Studies on hydraulic models

Due to the great size of the Tucuruí HPP project, hydraulic structures and equipment were studied in several reduced hydraulic models, which provided valuable inputs to the project.

The civil works were tested at the Saturnino Hydroelectric Laboratory of Brito–Hidroesb in Rio de Janeiro. These studies are presented in Chapter 10.

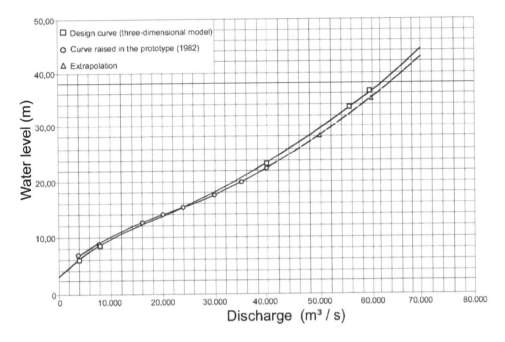

Figure 3.20 Culverts flow curves.

Source: (Carvalho, 1986).

Table 3.5 Culverts discharge coefficient (flow under pressure)

Q (m³/s)		*NA$_{mon}$ (m)	**H (m)	C
30.000	M	18,00	8,00	0,708
32.000	o	19,00	9,00	0,712
34.000	d	20,00	10,00	0,718
36.000	e	21,00	11,00	0,725
37.700	l	21,90	11,90	0,730
38.000		22,13	12,13	0,730
39.000		22,70	12,70	0,731
40.000		23,30	13,30	0,733
45.000		26,20	16,20	0,747
50.000		29,40	19,40	0,758
55.000		32,70	22,70	0,771
60.000		36,14	26,14	0,784
32.338	P	18,00	8,00	0,764
34.375	r	19,00	9,00	0,765
36.738	o	20,00	10,00	0,776
38.519	t	21,00	11,00	0,776
40.284	o	21,90	11,90	0,780
40.676	t	22,13	12,13	0,780
41.576	y	22,70	12,70	0,780
	p			
	e			

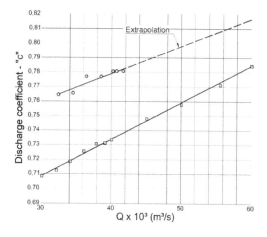

C values calculated from model measurements;
○ C values calculated from prototype measurements.

Figure 3.21 Culverts discharge curves.
Source: (Magela, 2015a).

Frame 3.2 WLs upstream of the culverts (flow under pressure)

Q (m³/s)	C extrapolated	H (m)	WL upstream = H + 10 (m)
50.000	0,797	17,56	27,56
60.000	0,816	24,12	34,12

Figure 3.22 Tucuruí HPP. Deviation through the culverts.
Source: (DGRB-CBDB – 2009).

3.6   Hydraulic design of other types of spillway – notes

3.6.1   Culvert spillway – bottom outlet

As stated in section 2.6, a culvert spillway (or bottom outlet) is used when a large discharge capacity is required in low head reservoirs (h < 25 ft = 7,6 m; DSD, pages 368 and 430). It is appropriate to refer to the work of Carter (1957) and Bodhaine (1968), US Geological

Observe:
- the size of the special blocks
(compare with the human scale);
- the slope (Δh) at the tip of the
landfill and the acceleration of
the flow.

Figure 3.23 Tucuruí HPP – third stage of river diversion; third stage pre-cofferdam.(A) Overview of the right bank channel. (B) Channel closure (final gap).

Survey, which are two treatises on the subject (available on the internet). Chow (1959) also deals with culverts (bottom holes). The following is a brief summary on the subject, with the aim of illustrating the theme for junior engineers.

A culvert is a unique type of constriction, and its entrance is a special kind of contraction. The characteristics of the flow are very complicated due to the many control variables, including the input geometry, the inclination, the size of the culvert, the roughness, the conditions of approach and restitution. The upstream flow is usually smooth and uniform. Within the culvert, however, the flow may be subcritical, critical, supercritical or fluvial, critical, torrential if the culvert is partially filled or under pressure. It works like a channel when it works partially full. The characteristics of the flow are illustrated in Figure 3.24:

(1)   the cross-section of the approach channel to an upstream distance of the inlet equivalent to a width of the culvert;
(2)   the entrance of the culvert;
(2–3) the culvert barrel;
(3)   the exit of the culvert;
(4)   the exit section of the tailwater (downstream *WL*).

The change in the waterline profile in the approach channel reflects the effect of the acceleration of the flow due to the contraction of the cross-section. The loss of energy near the entrance is related to the sudden contraction and subsequent expansion of the current inside the pipe; the entrance geometry has an important influence on this loss. The loss of energy by friction along the culvert is smaller, except in long and very inclined culverts.

The important aspects that control the stage-flow relationship in the approach section may be the occurrence of critical depth in the culvert, elevation of the downstream water level, the entrance geometry of the culvert, or a combination of these.

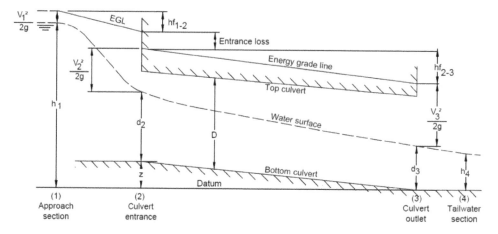

Figure 3.24 Definition sketch of culvert flow.

Source: (Bodhaine, 1968).

The peak flow through the culvert is determined by applying the continuity equation and the energy equation between the approach cross-section and a cross-section within the culvert. The location of the downstream cross-section depends on the state of the flow inside the culvert. For example, if critical flow occurs at the entrance, the elevation of the upstream water level is neither a function of the friction loss along the culvert nor of the upward water level elevation, and the terminal cross-section is located from the upstream side of the culvert (Frame 3.3, Type 1).

A culvert will work full when the outlet is submerged or when the outlet is not submerged, but the head is high (h_1) and the length (L) is long.

According to laboratory tests carried out by the US Geological Survey, by the Bureau of Public Roads and others, cited by Bodhaine (1968), the entrance of a normal pipe will not be submerged if the load is less than a critical value of the load, $h_1 - z = 1.5$ D. Laboratory tests have also indicated that a culvert, usually with a square top at the inlet, will not work full even if the inlet is below the reservoir water level when the outlet is free.

Under these conditions, the flow at the entrance of the culvert will shrink to a depth less than the height of the culvert in a manner very similar to the contraction of the flow in the form of a jet under a sluice gate. This high velocity jet will continue along the length of the culvert, but the velocity will be reduced by friction causing a gradual loss of charge. If the culvert is not long enough to allow expansion of the flow height, the culvert will never work full. Such a culvert is considered hydraulically short. Otherwise, the pipe is considered hydraulically long, and therefore will work fully as a conduit (Chow, 1959). For convenience in the calculations, the flow rates were classified as shown in Frame 3.3, which also contains the formulas for hydraulic design.

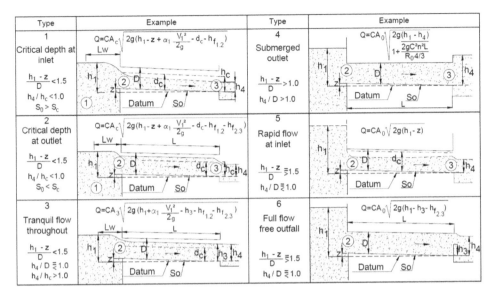

Frame 3.3 Classification of culvert flow and formulas for hydraulic design (Bodhaine, 1968).

The characteristics of the flow are given in Table 3.6. From this information, one can make the following general classification of the types of flow:

- if $h_4/D \leq 1,0$ and $(h_1 - z)/D < 1,5$ only flow types 1, 2 and 3 are possible;
- if $h_4/D > 1,0$, only flow type 4 is possible;
- if $h_4/D \leq 1,0$ and $(h_1 - z)/D \geq 1,5$, only flow types 5 and 6 are possible.

Table 3.6 Characteristics of flow types – (Bodhaine, 1968)

Type	Flow	Location of terminal section	Type of control	Slope of the culvert	$\dfrac{h_1 - z}{D}$	$\dfrac{h_4}{h_e}$	$\dfrac{h_4}{D}$
1	Parc full	Inlet	d_c	Steep	< 1,5	< 1,0	≤ 1,0
2	Parc full	Exit	d_c	Soft	< 1,5	< 1,0	≤ 1,0
3	Parc full	Exit	Backwater	Soft	< 1,5	> 1,0	≤ 1,0
4	Full	Exit	Backwater	Any	> 1,0	–	> 1,0
5	Parc full	Inlet	Inlet	Any	≥ 1,5	–	≤ 1,0
6	Full	Exit	Inlet and D	Any	≥ 1,5	–	≤ 1,0

Further information on the type of flow will require trial and error tests, as presented in detail by Bodhaine (1968). Due to the complexity of the subject, the proper determination of the characteristics of this flow should be done in studies in reduced model.

3.6.2 *Spillway in orifice*

As mentioned previously in section 2.8, spillway orifices used in the case of arch dams in narrow valleys launch a jet at a great distance from the dam and avoid erosion in the area near the dam. The jet range can be estimated as explained in section 5.2. At the same time, they ensure a reduction of erosive capacity as a result of the dispersion of the jet in the air (Lencastre, 1982). Among others, mention was made in section 2.8:

- the spillway of the Kariba HPP (Figures 5.18 to 5.21) and the spillway of the Cahora Bassa HPP (Figures 2.30 and 5.22 and 5.23), on the Zambesi river in Mozambique, which will be briefly discussed in Chapter 5 (Quintela and Cruz, 1982);
- the spillway at Mossyrock HPP, on the Cowlitz river in the Columbia river basin (Figure 2.31);
- the spillway Boundary dam (1,000 MW) on the Pend Oreille river, Figure 2.33.
- the spillway of the Sainte-Croix dam (Figure 3.24A), on the Verdon river, a tributary of the Durance river, Provence (France), completed in 1974, with a flow capacity of 1.100 m³/s, with a head on the 75 m sluice gates, shown in the following illustrations (Haüsler, CBDB/ICOLD, 1983).

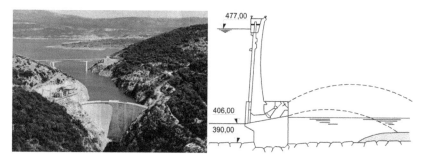

Figure 3.24A Sainte-Croix dam, Verdon river (France).

In designing these spillways, in addition to the geological and morphological characteristics of the site, the following points should be considered:

- discharge capacity: the alternative of having several orifices depending on the possibility of damage to one of the gates;
- structures in contact with the flow: the nature of the field of pressure acting on the structures and the risks of cavitation and vibration must be taken into account;
- regulating gates: experience shows good performance of the segment gates used to control the flows;
- positioning of the orifices: in addition to the factors already listed, the positioning is conditioned by the elevation of the upstream water level, the maximum permissible head values in the gates, the width of the valley, problems related to dam stability and vibration hazards, and cavitation; can be conditioned by the reach of the jets, in view of their erosive effects.

According to Lencastre (1972), the flow capacity can be estimated by the formula:

$$Q = cS\sqrt{2gh} \tag{3.5}$$

where:
C = coefficient of discharge;
S = area of the orifices (m²);
H = head on the center of the orifices (m).

3.6.3 *Labyrinth spillway*

As stated previously in section 2.9, a labyrinth spillway is characterized by a broken axis in the form of an accordion (U-shaped, V-shaped and trapezoidal) to increase the overall length of the crest for a given width available on the site in order to increase the discharge capacity.

These spillways are particularly advantageous in situations where it is necessary to discharge a large amount of water with a small charge on the crest. Additionally, they have much simpler operation and maintenance than a spillway with gates.

They have been designed and built for various flow capacities and are suitable for existing spillway rehabilitation projects. The structure is quite economical.

The total length of a labyrinth is on the order of three to five times the width available for the spillway in the site. Its capacity varies with the load but is usually twice as large as that of a standard spillway of the same width.

Variables that need to be considered in the design include labyrinth length and width, crest height, labyrinth angle, number of modules (cycles), and several other less important factors such as wall thickness, shape of the crest and the configuration of the vertex.

For details, see Taylor (1968), Hay and Taylor (1970), Lux III and Hinchliff (1985), USBR (1992), Tullis *et al.* (1995) and Mays (1999).

These spillways, in addition to increasing flow capacity for a given slope elevation, can be used to increase storage capacity while maintaining spillway capacity. Figure 3.25 and 3.26 show the layout of a labyrinth spillway of the Bonfante Hydroelectric Power Plant, close to BR-040 in the states of Minas Gerais and Rio de Janeiro division (Brazil).

Using the U, V and trapezoidal forms, the flow pattern is complex. Ideally, the discharge through the labyrinth should increase in direct proportion to the increase in crest length. However, this only occurs in the case of labyrinth spillways with low design head.

Qualitatively, when the head increases, the flow pattern passes, sequentially, through four basic phases: fully aerated (Figure 3.26E), partially aerated (Figure 3.26F), transition and no aeration (suppressed).

The fully aerated condition occurs at low heads when the flow falls freely over the entire spillway crest (Figure 3.26E). For this condition, the thickness of the nappe and the depth of the upstream water level (H_d in Figure 3.26A) do not affect the spillway capacity, and the labyrinth behaves almost ideally when compared to a linear spillway with the same cross-section.

As the head increases, the upstream water level increases, particularly between the nappe and the wall of the labyrinth.

Due to the convergence of the opposing nappes, the upstream water level uptake and the restricted area at the upstream vertices, aeration becomes difficult. This phenomenon interferes with the nappe, reduces the flow coefficient and defines the beginning of the partially aerated phase (Figure 3.26F).

Because of this, the upstream vertex flow becomes non-aerated (suppressed) and the volume of air under the nappe at the upstream vertex is insufficient to maintain aeration. A stable air bubble forms along each side of the wall and at the vertex downstream of the labyrinth.

In this last phase, the thickness of the nappe and the downstream water level obviously do not allow more aeration of the nappe, possibly resulting in a complete submergence, damaging the usefulness of the structure.

The complete submergence usually occurs when the head on the crest is greater than the height of the labyrinth, $Ho > P$ as defined in Figure 3.25A. For details see Lux III and Hinchliff (1985).

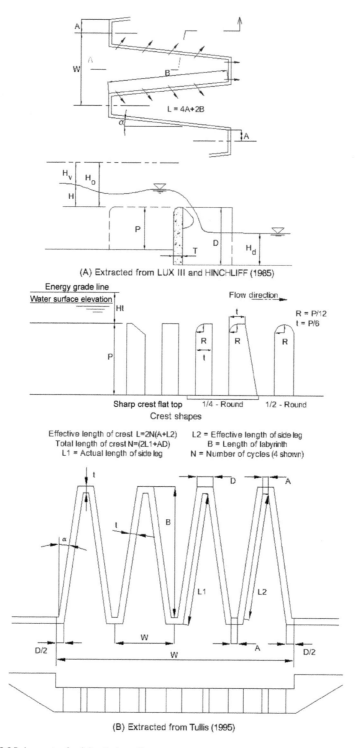

L = 4A+2B

(A) Extracted from LUX III and HINCHLIFF (1985)

Energy grade line
Water surface elevation

R = P/12
t = P/6

Sharp crest flat top   1/4 - Round   1/2 - Round
Crest shapes

Effective length of crest L=2N(A+L2)   L2 = Effective length of side leg
Total length of crest N=(2L1+AD)   B = Length of labyrinth
L1 = Actual length of side leg   N = Number of cycles (4 shown)

(B) Extracted from Tullis (1995)

Figure 3.25 Layout of a labyrinth spillway.

A – 2nd Phase of diversion.
Part of the labyrinth inside the cofferdam.

B – Plant – downstream view. Highway-040.
Bridge over the tailrace tunnel.

C – Downstream view detail.

D – Labyrinth spillway detail.

E – Fully aerated nappe (2008).

F – Nappe in the transition phase (2008).

Figure 3.26 Bonfante HPP. Labyrinth spillway.

The discharge capacity can be estimated using the following equation.

$$Q = \frac{2}{3} C_d L \sqrt{2g} H_t^{3/2}$$

(3.6)

where:
Q = discharge capacity (m³/s);
C_d = 0,30, discharge coefficient, from Figure 3.27 with α = 8° and H_t/P = 0,8;
L = total spill crest length (m);
H_t = head (m).

Set vertex A and determine the number of N modules of the labyrinth using $L = 2N(A + L_2)$, where L_2 is dimensioned to produce an integral N number.

Another attempt should be made assuming α different from 8° and varying H_t/P ($\leq$0.9) to check the project's economics.

After selecting the size of the labyrinth, the details are: $t = P/6$; $R = P/2$: $A = 1$ to 2 t. The crest shape is quarter-round.

The width of the spillway downstream, W, is equal to the width of the labyrinth. The slope of the chute should be supercritical for the entire flow range.

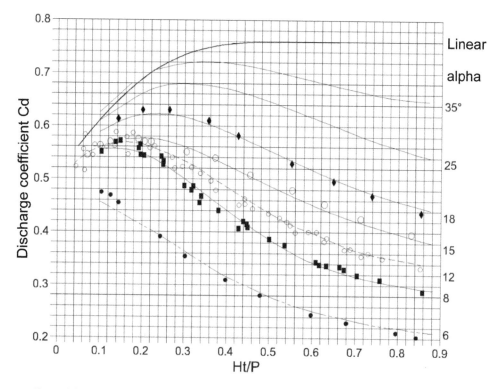

Figure 3.27 Discharge coefficients.

Source: (Tullis et al., 1995)

Table 3.7 Labyrinth spillway design

Parameter	Symb.	Value	Un	Source/Equation/Notes
(a) Data				
Maximum discharge	Q_{max}		m³/s	Given
WL max	NA_{res}		m	Given
Approach channel El.	EL_{ca}		m	Given
Crest El.	EL_c		m	Given
Total head	H_t		m	$H_t = WL_{res} - EL_c - loss$
(b) Data taken				
Entrance loss	Loss		m	Estimated
Nº of modules	N		–	Select keep w/P~3 to 4
Cres height	P		m	Fix it P ~ 1,4 H_t
Angle of side flaps	A		°C	Normally 8° – 16°
(c) Calculated data				
Thickness wall	T		m	t = P/6
Inside vertex width	A		m	Select between t and 2t
External vertex width	D		m	D = A + 2 t tan (45 – α/2)
Head/Crest height	H_t/P		–	-
Discharge coefficient	C_d		–	From Figure 3.27
Effective crest length	L		m	$1,5\ Q_{max}/[(C_{dH_t^{0,5}})(2g)^{0,5}]$
Length of the module	B		m	B = [L/(2N) + t tan (45 – α/2)] cos α + t
Length of the flap	L_1		m	L_1 = (B – t)/cos α
Length effective of the flap	L_2		m	$L_2 = L_1 - $ t tan (45 – α/2)
Total wall length	L_3		m	L_3 = N (2L_1 + D + A)
Distance between modules	w		m	w = 2L_1 sin α + A + D
Labyrinth width	W		m	W = N w
Linear length of the labyrinth	–		m	$1,5\ Q_{max}/[(C_{dH_t^{0,5}})(2g)^{0,5}]$
Dist. between modules /P	w/P		–	Usually between 3 and 4
(d) Volume of concrete				
Wall	–		m³	Vol = L_3 P t
Slab**	–		m³	Vol = W B t
Total	–		m³	

* C_d for linear spillway;
** For the concrete volume of the slab, it is assumed that the slab thickness is equal to that of the wall.

Table 3.7, extracted from Tullis *et al.* (1995), shows a worksheet with the hydraulic design guidelines.

3.6.4 Stepped spillway

As mentioned previously in section 2.11, with the introduction of roll compacted (CCR) dams, it became convenient to leave steps on the downstream chute of the spillway. The steps break the flow dissipating the energy, producing a skimming flow (white waters) and reducing the excess to be dissipated in the basin. This bring economic benefits to the project by reducing the length of the stilling basin.

Figure 3.28 shows the solution adopted for the Dona Francisca HPP (125 MW), on the Jacuí river in the State of Rio Grande do Sul in the Southern Region of Brazil, for a discharge of 10.600 m³/s, with a width of 335 m and a basin with a length of 42,38 m. It was possible to reduce the length of the basin by 15%.

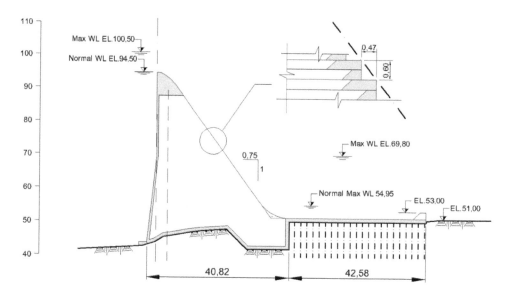

Figure 3.28 Dona Francisca hydroelectric power plant. Stepped spillway.
Source: (CBDB, MDB, 2000).

Project procedure:

- design the crest without control (shape and length) as described at the beginning of Chapter 2;
- design the height of the step in such a way that the ratio between critical depth at the crest and the height of the step is $Yc/h < 4$, being $Y_C = \sqrt[3]{q^2/g}$;
- determine the loss of head along the steps, from the reservoir to the basin, based on Figures 3.29 and 3.30 ($H = N \times h$, where N = number of steps);
- hydraulic design of the stilling basin as described in Chapter 4.

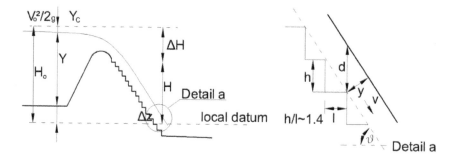

Figure 3.29 Stepped spillway. Notations.

For details on this type of spillway, consult, among others: "An Investigation on Stepped Spillways" (CBDB/LBS, 2002), where an extensive list of references on the subject can be found, such as CIRIA (1978), Tozzi (1992), Chanson (2001) and USBR (2002). About the project Dona Francisca, consult Sobrinho (2000).

Figure 3.30 Relation between the loss of head and critical depth for a stepped spillway.
Source: (Christodoulou, 1993, USBR, 2002, 2015).

Figure 3.31 Flow pattern in the laboratory, for project discharge.
Source: (www.vaw.ethz.ch).

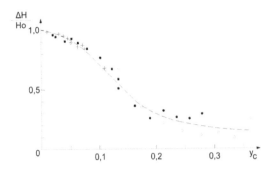

Figure 3.32 Dona Francisca HPP (Brazil).

3.6.5 *Morning glory spillway*

The morning glory spillway is usually used in conjunction with a tunnel when the water intake is a vertical shaft, as previously mentioned in section 2.4. In the design, this is used in USBR DSD (1974), section 212. Oliveira (1986) and Lemos (1986) are also cited.

The flow enters through the periphery of the shaft and the capacity of the crest is high. Usually, gates are not used because of access conditions and cost considerations.

Figure 3.34 shows the elements and shape of the nappe and profile.

Figure 3.33 Morning glory spillway of Monticello dam (Napa, CA-USA). $D = 22$ m. $Q = 1.370$ m^3/s. (see photos, section 2.4).

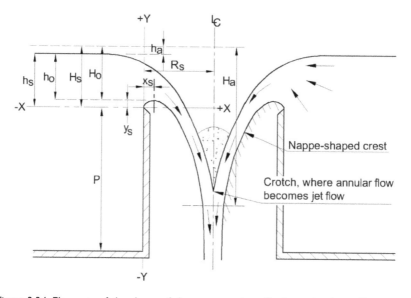

Figure 3.34 Elements of the shape of the nappe and profile for a circular spillway.
Source: (DSD, 1974).

Figure 3.35, from Linsley and Franzini (1955), shows in a simple and very objective manner a typical tulip spillway, the flow conditions and the simplified discharge curve with its characteristics.

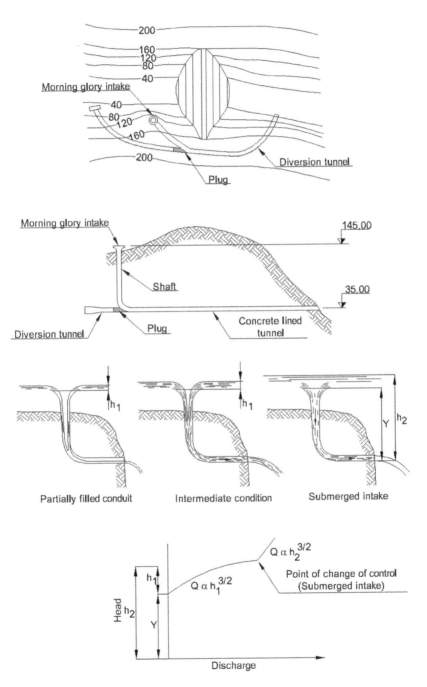

Figure 3.35 Nature of flow and discharge characteristics.

Source: (Linsley and Franzini, 1955).

Discharge

The flow conditions and the discharge curve are detailed in Figure 3.36, which was extracted from DSD (1974).

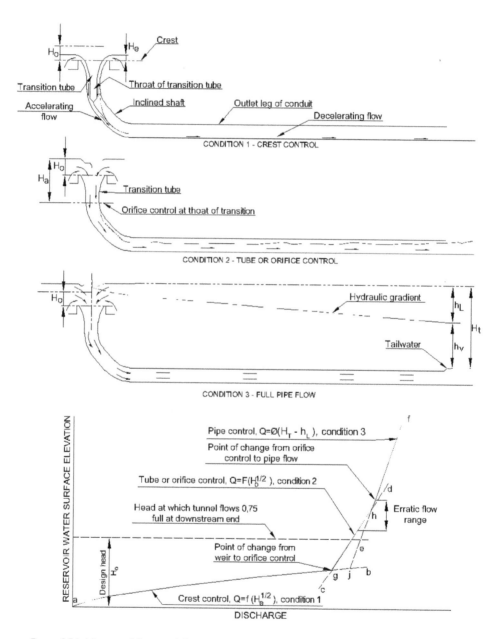

Figure 3.36 Nature of flow and discharge characteristics.

Source: (DSD, 1974, Illustration 281).

The characteristics of the flow in a morning glory spillway vary according to the proportions of the various elements: crest diameter, transition and throat size.

By changing the crest diameter, the "ab" curve will be changed so that the ordinate "g" on the "cd" curve may be higher or lower.

For a larger diameter, the discharges will be higher for lower loads. The transition will completely fill and control of the tube will occur with the slightest head in the crest.

Similarly, by changing the size of the throat, the position of the curve "cd" will be changed and will indicate the heads above which the pipe control will prevail.

If the transition is made such that the curve "cd" is moved to match or position to the right of the "j" point, the control will move directly from the crest to the end of the conduit.

The details of the hydraulic characteristics of the flow are presented in DSD (1974). A summary is given here.

As shown in the discharge curve shown in Figure 3.36, the nature of the flow over a morning glory spillway and the discharge characteristics can be summarized as follows:

- crest control will prevail for heads between ordinates "a" and "g" (condition 1);
- orifice or tube control for heads between ordinates "g" and "h" (condition 2); and,
- the spillway conduit will flow full for heads higher than the ordinate "h" (condition 3).

For small heads the flow in the tulip is governed by the discharge characteristics of the crest. The vertical transition below the crest will be flowing partially full, and the flow will stick to the sides of the shaft. As the discharge on the crest increases, the nappe will become thick and eventually converge in a solid vertical jet at the point called the fork (crotch).

After formation of the solid jet, a "boil zone" (boil) will occupy the region above the fork. Both the fork and top of the boiling zone become progressively higher with the higher discharges. For larger heads, the fork and the boiling zone may almost overflow, showing only a gentle depression and a vortex on the surface.

Until the solid jet formation, free discharge prevails at the spillway. After the fork and the formation of the boiling zone, the submergence begins to affect the flow through the spillway until the crest is flooded.

In this situation, the flow is then governed either by the nature of the contracted jet, which is formed by the flood of the inlet, or by the shape and size of the vertical transition, if it does not conform to the jet shape. The action of the vortex should be minimized to maintain the flow converging in the shaft, using, for example, guide pillars.

If the crest profile and transition conform to the shape of the lower nappe of the jet flowing above the sharp-crested circular weir, the discharge can be expressed by the equation:

$$Q = CLH^{3/2} \tag{3.7}$$

where:
H is the head on the spillway;
L is the length of the circle along the periphery at apex to crest.

The value of C will vary according to the values of L and H. If L is measured at the outer periphery of the spill nappe (according to the coordinate system shown in Figure 3.34) and if the head is measured at the apex of the leaf, equation 3.6 can be rewritten:

$$Q = C_0(2\pi R_S)H_0^{3/2} \qquad\qquad (3.7a)$$

It will be apparent that the flow coefficients for circular crests differ from those for straight crests because of the effects of the submergence and the incident back pressure for the convergence flow union.

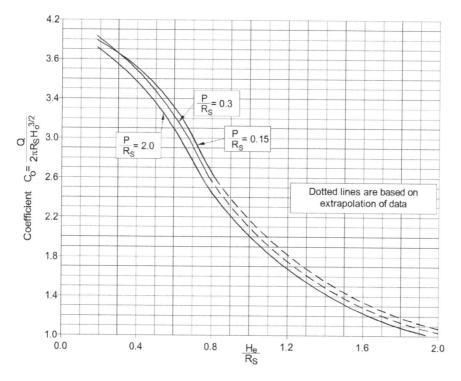

Figure 3.37 Relations of the coefficient of circular crest C_o for H_o/R_s for different depths of approach (aerated nappes).

Source: (DSD, 1974, Illustration 283).

The coefficient C_o must be related to H_o and R_s and must be expressed in terms of H_o/R_s (Figure 3.37).

These coefficients are valid only if the crest profile and the shape of the transition adjust to that of the jet flowing over the circular sill at the head H_o and if aeration is sufficient so that there are no negative pressures along the contact of the lower nappe.

When the crest design and transition shape conform to the nappe profile for a given H_o on the crest, the free flow prevails for $H_o/R_s \sim 0.45$, and the weir governs.

When the relation is $\frac{H_0}{R_S} > 0,45$, the spillway submerges, and the flow showing these characteristics is the control condition.

When $\frac{H_0}{R_S} \tilde{} 1,0$, the spillway is completely submerged. For this condition and higher stages of $\frac{H_0}{R_S}$, the phenomenon is that of flow through orifice.

It should be noted that for most flow conditions over a circular spillway the discharge coefficient increases with a reduction of approach depth (P), while the opposite is true for straight spillways.

For both spillways, a reduction of the depth of the approach flow decreases the vertical component of the velocity and consequently suppresses the contraction of the nappe.

However, for circular spillways the submergence effect is reduced because of the depressed surface of the upper nappe, giving the jet a rapid downward thrust, which lowers the crotch position and increases the discharge.

Flow coefficients for partial loads at the crest, H_e, can be determined using the graph of Figure 3.38 to construct a spillway flow capacity curve.

The designer should take precautions in applying this criterion, since negative pressures or submergence effects may change the flow conditions differently for various shaped profiles. This criterion, however, should not be adopted for flow conditions where $H_e/R_s > 0,4$.

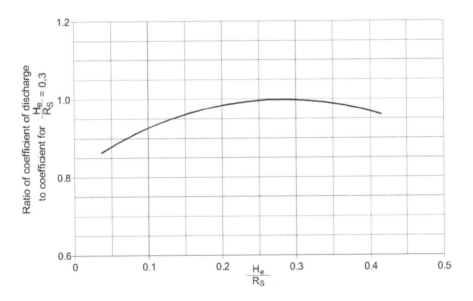

Figure 3.38 Flow coefficients of the circular crest for other heads beyond the design head.

Source: (DSD, 1974, Illustration 284).

Crest profiles

The values of the coordinates to define the shape of the lower nappe on the crest of the tulip, for various conditions of $\frac{P}{R_S}$ and $\frac{H_S}{R_S}$, are shown in Tables 3.8, 3.9 and 3.10 near the end of this chapter. These data are based on experiments performed in the USBR (Wagner, 1956). The relationships of H_s to H_o are shown in Figure 3.39.

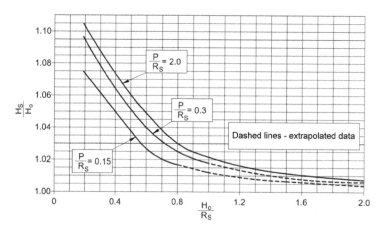

Figure 3.39 Relations of H_s/H_o to H_o/R_s for circular crest spillways.
Source: (DSD, 1974, Illustration 285).

Typical upper and lower nappe profiles for various values of $\frac{H_S}{R_S}$ are plotted in Figure 3.40 in terms of $\frac{x}{H_S}$ and $\frac{y}{H_S}$ for the condition of $\frac{P}{R_S} = 2, 0$ (aerated nappes and negligible approach velocity).

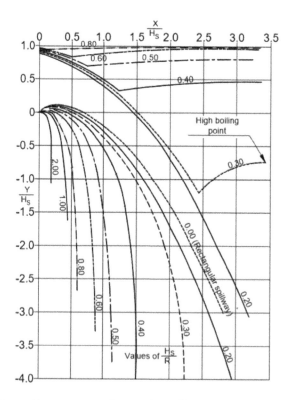

Figure 3.40 Top and bottom nappe for circular spillway.
Source: (DSD, 1974, Illustration 286)

Figure 3.41 shows typical lower nappe profiles for various values of H_s and for a given value of R_s. In contrast to straight spillways, where the nappe jumps farther from the crest as the head increases, for a circular crest the bottom nappe jumps far only in the high-point region of the track, and only for $H_s/R > 0.5$.

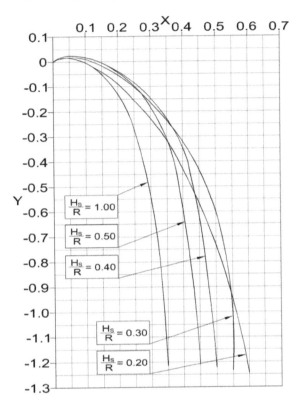

Figure 3.41 Comparison of lower nappe forms for spillway of different heads, for a given value of R_s.

Source: (DSD, 1974, Illustration 287)

Figure 3.42 shows the required increase in radius to minimize negative crest pressures. The shape of the crest for the increased radius is then based on a relation $H_S/R_S = 0,3$.

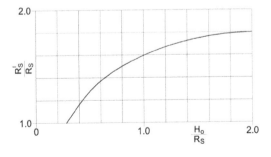

Figure 3.42 Required increase of radius of the circular crest to minimize negative pressures along the crest.

Source: (DSD, 1974, Illustration 288).

Transition project

The jet diameter coming out of a horizontal orifice can be determined to any point below the surface of the water assuming that the continuity equation, $Q = v\,A$, is valid and if friction and other losses are neglected.

For a circular jet the area is $A = \pi R^2$. Therefore $v\,A = \pi R^2 \sqrt{2gh}$. Resolving to R, it has been:

$$R = \frac{Q_a^{1/2}}{5H_a^{1/4}} \tag{3.8}$$

where:
H_a = the difference between the *WL* elevation of the reservoir and the elevation under consideration.

The jet diameter therefore decreases with the distance from the vertical free fall to the normal design applications.

Assuming that the total loss of charge of the order of 0,1 H_a, the equation can be rewritten:

$$R = 0,204 \frac{Q_a^{1/2}}{H_a^{1/4}} \tag{3.9}$$

$$R = 0,275 \frac{Q_a^{1/2}}{H_a^{1/4}}, \quad \text{for metric units} \tag{3.9a}$$

Since this equation is for the jet shape, its use to determine the shape of the shaft will result in a minimum size that will accommodate the flow without restrictions and without the development of pressures along the side of the shaft. A typical profile of a shaft determined from this equation is shown by dash-dotted lines "abc" in Figure 3.43.

If the shaft profile, "abc," is raised above the "b" point, as shown by the dotted line "db," in section A-A, the flow will be under pressure; below cross-section A-A the free jet profile should follow the lines "bc."

Aeration is required in the control or by the introduction of air for a rapid widening of the well or the installation of a baffle (deflector) to ensure free flow below the control in cross-section A-A.

The sizes of elbows, chutes and slopes shall be such that the free flow is maintained below the control point. Inadequate aeration at the control point can introduce cavitation risks.

For the "abe" profile, established for a given head, the control must remain in section A-A for any greater head such that above the section the flow under pressure will prevail.

The flow below cross-section A-A must be kept free. If the "dbe" profile is adopted, once a head is reached to run the full shaft at point "b," cross-section AA should be the control point, and the pressure flow above the control will prevail for that and other larger heads.

For submerged crest flow, the corresponding nappe shape as determined in the previous cross-section for a given H_0 will be such that along the low levels it will closely follow the profile determined, using Equation 3.8 if H_e approaches H_0.

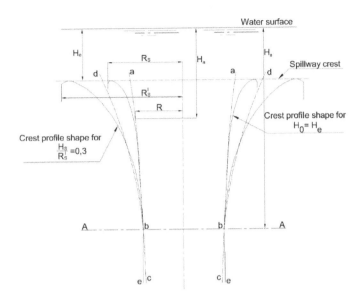

Figure 3.43 Comparison of shaft profile profiles for various flow conditions.
Source: (DSD, 1974, Illustration 289).

Conduit design

If, for a given discharge, the conduit of a morning glory spillway is to flow full below the transition without being under pressure, the required size of the shaft and the length of the conduit should vary according to the net head available along the length.

Since the inclination of the energy gradient, which is dictated by the charge losses, is smoother than the incline of the conduit, the flow will accelerate and the conduit will decrease.

As the slope of the conduit becomes softer than the hydraulic gradient, the flow will decelerate and the conduit will increase in size.

All conduit points will act simultaneously to control the flow discharge. For heads in excess of that used to determine the size of the conduit, the flow will be under pressure with the control at the downstream outlet; for smaller heads, the conduit will work partially full throughout its length and the control will remain in the upstream transition.

In Figure 3.35, the head on which the conduit works exactly full is represented by the "h" point. For heads above point "h" the conduit will work full; for smaller heads, the conduit will work partially filled, with control being dictated by the transition project.

Because of the impracticability of constructing a conduit with variable diameter, it is routinely made with the constant diameter below the transition point. Therefore, the conduit will have an excess of area from this point.

If the atmospheric pressure can be maintained along the partially filled conduit, it will continue to work even if the downstream outlet is drowned. Larger discharges will not progressively alter the working full section at the top of the conduit, but the full flow conditions under pressure will occupy larger stretches in the downstream stretch.

In the discharge represented by the "h" point in Figure 3.35, the full flow condition will move back to the transition zone and the conduit will work full over its entire length.

Table 3.8 Coordinates of the lower water table for different values of H_s/R when $P/R = 2$

(negligible approach velocity and aerial nappe)

H_s/R	0,00	0,10*	0,20	0,25	0,30	0,35	0,40	0,45	0,50	0,60	0,80	1,00	1,20	1,50	2,00
X/H_s	Y/H_s For the portion of the profile above the crest of the spillway														
0,00	0,0000	0,0000	0,0000	0,0000	0,0000	0,0000	0,0000	0,0000	0,0000	0,0000	0,0000	0,0000	0,0000	0,0000	0,0000
0,01	0,0150	0,0145	0,0133	0,0130	0,0128	0,0125	0,0122	0,0119	0,0116	0,0112	0,0104	0,0095	0,0086	0,0077	0,0070
0,02	0,0280	0,0265	0,0250	0,0243	0,0236	0,0231	0,0225	0,0220	0,0213	0,0202	0,0180	0,0159	0,0140	0,0115	0,0090
0,03	0,0395	0,0365	0,0350	0,0337	0,0327	0,0317	0,0308	0,0299	0,0289	0,0270	0,0231	0,0198	0,0168	0,0126	0,0090
0,04	0,0490	0,0460	0,0435	0,0417	0,0403	0,0389	0,0377	0,0363	0,0351	0,0324	0,0268	0,0220	0,0176	0,0117	0,0085
0,05	0,0575	0,0535	0,0506	0,0487	0,0471	0,0454	0,0436	0,0420	0,0402	0,0368	0,0292	0,0226	0,0168	0,0092	0,0050
0,06	0,0650	0,0605	0,0570	0,0550	0,0531	0,0510	0,0489	0,047	0,0448	0,0404	0,0305	0,0220	0,0147	0,0053	
0,07	0,0710	0,0665	0,0627	0,0605	0,0584	0,0560	0,0537	0,0514	0,0514	0,0487	0,0308	0,0201	0,0114	0,0001	
0,08	0,0765	0,0710	0,0677	0,0655	0,0630	0,0603	0,0578	0,0550	0,0521	0,0455	0,0301	0,0172	0,0070		
0,09	0,0820	0,0765	0,0722	0,0696	0,0670	0,0640	0,0613	0,0581	0,0549	0,0471	0,0287	0,0135	0,0018		
0,10	0,0860	0,0810	0,0762	0,0734	0,0705	0,0672	0,0642	0,0606	0,0570	0,0482	0,0264	0,089			
0,12	0,0940	0,0880	0,0826	0,0790	0,0758	0,0720	0,0683	0,0640	0,0596	0,0483	0,0195				
0,14	0,1000	0,0935	0,0872	0,0829	0,0792	0,0750	0,05	0,0654	0,0599	0,0460	0,0101				
0,16	0,1045	0,0980	0,0905	0,0855	0,0812	0,0765	0,0710	0,0651	0,0585	0,0418					
0,18	0,1080	0,1010	0,0927	0,0872	0,0820	0,0766	0,0705	0,0637	0,0559	0,0361					
0,20	0,1105	0,1025	0,938	0,0877	0,0819	0,0756	0,0688	0,0611	0,0521	0,0292					
0,25	0,1120	0,1035	0,0926	0,0850	0,0773	0,0683	0,0596	0,0495	0,0380	0,0068					
0,30	0,1105	0,1000	0,0850	0,0764	0,0668	0,0559	0,0446	0,0327	0,0174						
0,35	0,1060	0,0930	0,0750	0,0650	0,0540	0,0410	0,0280	0,0125							
0,40	0,0970	0,0830	0,0620	0,0500	0,0365	0,0220	0,0060								
0,45	0,0845	0,0700	0,0450	0,0310	0,0170	0,000									
0,50	0,0700	0,0520	0,0250	0,0100											
0,55	0,0520	0,0320	0,0020												
0,60	0,0320	0,0080													
0,65	0,0090														
Y/H_s	x/H_s For the portion of the profile below the crest of the spillway														
0,00	0,668	0,615	0,554	0,520	0,487	0,450	0,413	0,376	0,334	0,262	0,158	0,116	0,093	0,070	0,048
-0,02	0,705	0652	0,592	0,560	0,526	0,488	0,452	0,414	0,369	0,293	0,185	0,145	0,120	0,096	0,074
-0,04	0,742	0,688	0,627	0,596	0,563	0,524	,487	0,448	0,400	0,320	0,212	0,165	0,140	0,115	0,088

(Continued)

Table 3.8 (Continued)

(negligible approach velocity and aerial nappe)

H_s/R	0,00	0,10*	0,20	0,25	0,30	0,35	0,40	0,45	0,50	0,60	0,80	1,00	1,20	1,50	2,00
-0,06	0,777	0,720	0,660	0,630	0,596	0,557	0,519	0,478	0,428	0,342	0,232	0,182	0,155	0,129	0,100
-0,08	0,808	0,752	0,692	0,662	0,628	0,589	0,549	0,506	0,454	0,363	0,250	0,197	0,169	0,140	0,110
-0,10	0,838	0,784	0,722	0,692	0,657	0618	0,577	0,532	0,478	0,381	0,266	0,210	0,180	0,150	0,118
-0,15	0,913	0,857	0,793	0,762	0,725	0,686	0,641	0,589	0,531	0,423	0,299	0,238	0,204	0,170	0,132
-0,20	0,978	0,925	0,860	0,826	0,790	0,745	0,698	0,640	0,575	0,459	0,326	0,260	0,224	0,184	0,144
-0,25	1,040	0,985	0,919	0,883	0,847	0,801	0,750	0,683	0,613	0,490	0,348	0,280	0,239	0,196	0,153
-0,30	1,100	1,043	0,976	0,941	0,900	0,852	0,797	0,722	0,648	0,518	0,368	0,296	0,251	0,206	0,160
-0,40	1,207	1,150	1,079	1,041	1,000	0,944	0,880	0,791	0,706	0,562	0,400	0,322	0,271	0,220	0,168
-0,50	1,308	1,246	1,172	1,131	1,087	1,027	0,951	0,849	0,753	0,598	0,427	0,342	0,287	0,232	0,173
-0,60	1,397	1,335	1,260	1,215	1,167	1,102	1,012	0,898	0,793	0,627	0,449	0,359	0,300	0,240	0,179
-0,80	1,563	1,500	1,422	1,369	1,312	1,231	1,112	0,974	0,854	0,673	0,482	0,384	0,320	0,253	0,184
-1,00	1,713	1,646	1,564	1,508	1,440	1,337	1,189	1,030	0,899	0,710	0,508	0,402	0,332	0,260	0,188
-1,20	1,846	1,780	1,691	1,635	1,553	1,422	1,248	1,074	0,933	0,739	0,528	0,417	0,340	0,266	
-1,40	1,970	1,903	1,808	1,748	1,653	1,492	1,293	1,108	0,963	0,760	0,542	0,423	0,344		
-1,60	2,085	2,020	1,918	1,855	1,742	1,548	1,330	1,133	0,988	0,780	0,553	0,430			
-1,80	2,196	2,130	2,024	1,957	1,821	1,591	1,358	1,158	1,008	0,797	0,563	0,433			
-2,00	2,302	2,234	2,126	2,053	1,891	1,630	1,381	1,180	1,025	0,810	0,572				
-2,50	2,557	2,475	2,354	2,266	2,027	1,701	1,430	1,221	1,059	0,838	0,588				
-3,00	2,778	2,700	2,559	2,428	2,119	1,748	1,468	1,252	1,086	0,853					
-3,50	—	2,916	2,749	2,541	2,171	1,777	1,489	1,267	1,102						
-4,00	—	3,114	2,914	2,620	2,201	1,796	1,500	1,280							
-4,50	—	3,306	3,053	2,682	2,220	1,806	1,509								
-5,00	—	3,488	3,178	2,734	2,227	1,811									
-5,50	—	3,653	3,294	2,779	2,229										
-6,00	—	3,820	3,405	2,812	2,232										

* The tabulation for $H_s/R=0,10$ was obtained by interpolation between $H_s/R=0$ and 0,20. Tabel 25, DSD (1974).

Table 3.9 Coordinates of the bottom water different values of Hs/R when P/R = 0,30

Hs/R	0,20	0,25	0,30	0,35	0,40	0,45	0,50	0,60	0,80
x/Hs	Y/Hs For the portion of the profile above the crest of the spillway								
0,00	0,0000	0,0000	0,0000	0,0000	0,0000	0,0000	0,0000	0,0000	0,0000
0,01	0,0130	0,0130	0,0130	0,0125	0,0120	0,0120	0,0115	0,0110	0,0100
0,02	0,0245	0,0242	0,0240	0,0235	0,0225	0,0210	0,0195	0,0180	0,0170
0,03	0,0340	0,0335	0,0330	0,0320	0,0300	0,0290	0,0270	0,0240	0,0210
0,04	0,0415	0,0411	0,0390	0,0380	0,0365	0,0350	0,0320	0,0285	0,0240
0,05	0,0495	0,0470	0,0455	0,0440	0,0420	0,0395	0,0370	0,0325	0,0245
0,06	0,0560	0,0530	0,0505	0,0490	0,0460	0,0440	0,0405	0,0350	0,0250
0,07	0,0610	0,0575	0,0550	0,0530	0,0500	0,0470	0,0440	0,0370	0,0245
0,08	0,0660	0,0620	0,0590	0,0565	0,0530	0,0500	0,0460	0,0385	0,0235
0,09	0,0705	0,0660	0,0625	0,0595	0,0550	0,0520	0,0480	0,0390	0,0215
0,10	0,0740	0,0690	0,0660	0,0620	0,075	0,0540	0,0500	0,0395	0,0190
0,12	0,0800	0,0750	0,0705	0,0650	0,0600	0,0560	0,0510	0,0380	0,0120
0,14	0,0840	0,0790	0,0735	0,0670	0,0615	0,0560	0,0515	0,0355	0,0020
0,16	0,0870	0,0810	0,0750	0,0675	0,0610	0,0550	0,0500	0,0310	
0,18	0,0885	0,0820	0,0755	0,0675	0,0600	0,0535	0,0475	0,0250	
0,20	0,0885	0,0820	0,0745	0,0660	0,0575	0,0505	0,0435	0,0180	
0,25	0,0855	0,0765	0,0685	0,0590	0,0480	0,0390	0,0270		
0,30	0,0780	0,0670	0,0580	0,0460	0,0340	0,0220	0,0050		
0,35	0,0660	0,0540	0,0425	0,0295	0,0150				
0,40	0,0495	0,0370	0,0240	0,0100					
0,45	0,0300	0,0170	0,0025						
0,50	0,090	−,0060							
0,55									
Y/Hs	X/Hs For the portion of the profile below the crest of the spillway								
0,00	0,519	0,488	0,455	0,422	0,384	0,349	0,310	0,238	0,144
−0,02	0,560	0,528	0,495	0,462	0,423	0,387	0,345	0,272	0,174
−0,04	0,598	0,566	0,532	0,498	0,458	0,420	0,376	0,300	0,198
−0,06	0,632	0,601	0,567	0,532	0,491	0,451	0,406	0,324	0,220
−0,08	0,664	0,634	0,600	0,564	0,522	0,480	0,432	0,348	0,238
−0,10	0,693	0,664	0,631	0,594	0,552	0,508	0,456	0,368	0,254
−0,15	0,760	0,734	0,701	0,661	0,618	0,569	0,510	0,412	0,290
−0,20	0,831	0,799	0,763	0,723	0,677	0,622	0,558	0,451	0,317
−0,25	0,893	0,860	0,826	0,781	0,729	0,667	0,599	0,483	0,341
−0,30	0,953	0,918	0,880	0,832	0,779	0,708	0,634	0,510	0,362
−0,40	1,060	1,024	0,981	0,932	0,867	0,780	0,692	0,556	0,396
−0,50	1,156	1,119	1,072	1,020	0,938	0,841	0,745	0,595	0,424
−0,60	1,242	1,203	1,153	1,098	1,000	0,891	0,780	0,627	0,446
−0,80	1,403	1,359	1,0301	1,227	1,101	0,970	0,845	0,672	0,478
−1,00	1,549	1,498	1,430	1,333	1,180	1,028	0,892	0,707	0,504
−1,20	1,680	1,622	1,543	1,419	1,240	1,070	0,930	0,733	0,524
−1,40	1,800	1,739	1,647	1,489	1,287	1,106	0,959	0,757	0,540
−1,60	1,912	1,849	1,740	1,546	1,323	1,131	0,983	0,778	0,551
−1,80	2,018	1,951	1,821	1,590	1,353	1,155	1,005	0,797	0,560
−2,00	2,120	2,049	1,892	1,627	1,380	1,175	1,022	0,810	0,569
−2,50	2,351	2,261	2,027	1,697	1,428	1,218	1,059	0,837	
−3,00	2,557	2,423	2,113	1,747	1,464	1,247	1,081	0,852	
−3,50	2,748	2,536	2,167	1,778	1,489	1,263	1,099		
−4,00	2,911	2,617	2,200	1,796	1,499	1,274			
−4,50	3,052	2,677	2,217	1,805	1,507				
−5,00	3,173	2,731	2,223	1,810					
−5,50	3,290	2,773	2,228						
−6,00	3,400	2,808							

Table 26, DSD (1974)

Table 3.10 Coordinates of the bottom water different values of Hs/R when P/R = 0,15

Hs/R	0,20	0,25	0,30	0,35	0,40	0,45	0,50	0,60	0,80
X/Hs	Y/Hs For the portion of the profile above the crest of the spillway								
0,00	0,0000	0,0000	0,0000	0,0000	0,0000	0,0000	0,0000	0,0000	0,0000
0,01	0,0120	0,0120	0,0115	0,0115	0,0110	0,0110	0,0105	0,0100	0,0090
0,02	0,0210	0,0200	0,0195	0,0190	0,0185	0,0180	0,0170	0,0160	0,0140
0,03	0,0285	0,0270	0,0265	0,0260	0,0250	0,0235	0,0225	0,0200	0,0165
0,04	0,0345	0,0335	0,0325	0,0310	0,0300	0,0285	0,0265	0,0230	0,0170
0,05	0,0405	0,0385	0,0375	0,0360	0,0345	0,0320	0,0300	0,0250	0,0170
0,06	0,0450	0,0430	0,0420	0,0400	0,0380	0,0355	0,0330	0,0265	0,0165
0,07	0,0495	0,0470	0,0455	0,0430	0,0410	0,0380	0,0350	0,0270	0,0150
0,08	0,0525	0,0500	0,0485	0,0460	0,0435	0,0400	0,0365	0,0270	0,0130
0,09	0,0560	0,0530	0,0510	0,0480	0,0455	0,0420	0,0370	0,0265	0,0100
0,10	0,0590	0,0560	0,0535	0,0500	0,0465	0,0425	0,0375	0,0255	0,0065
0,12	0,0630	0,0600	0,0570	0,0520	0,0480	0,0435	0,0365	0,0220	
0,14	0,0660	0,0620	0,0585	0,0525	0,0475	0,0425	0,0345	0,0175	
0,16	0,0670	0,0635	0,0590	0,0520	0,0460	0,0400	0,0305	0,0110	
0,18	0,0675	0,0635	0,0580	0,0500	0,0435	0,0365	0,0260	0,0040	
0,20	0,0670	0,0625	0,0560	0,0465	0,0395	0,0320	0,0200		
0,25	0,0615	0,0560	0,0470	0,0360	0,0265	0,0160	0,0015		
0,30	0,0520	0,0440	0,0330	0,0210	0,0100				
0,35	0,0380	0,0285	0,0165	0,0030					
0,40	0,0210	0,0090							
0,45	0,015								
0,50									
0,55									
Y/Hs	X/Hs For the portion of the profile below the crest of the spillway								
0,00	0,454	0,422	0,392	0,358	0,325	0,288	0,253	0,189	0,116
−0,02	0,499	0,467	0,437	0,404	0,369	0,330	0,292	0,228	0,149
−0,04	0,540	0,509	0,478	0,444	0,407	0,368	0,328	0,259	0,174
−0,06	0,579	0,547	0,516	0,482	0,443	0,402	0,358	0,286	0,195
−0,08	0,615	0,583	0,550	0,516	0,476	0,434	0,386	0,310	0,213
−0,10	0,650	0,616	0,584	0,547	0,506	0,462	0,412	0,331	0,228
−0,15	0,726	0,691	0,660	0,620	0,577	0,526	0,468	0,376	0,263
−0,20	0,795	0,760	0,729	0,685	0,639	0,580	0,516	0,413	0,293
−0,25	0,862	0,827	0,790	0,743	0,692	0,627	0,557	0,445	0,319
−0,30	0,922	0,883	0,843	0,797	0,741	0,671	0,594	0,474	0,342
−0,40	1,029	0,988	0,947	0,893	0,828	0,749	0,656	0,523	0,381
−0,50	1,128	1,086	1,040	0,980	0,902	0,816	0,710	0,567	0,413
−0,60	1,220	1,177	1,129	1,061	0,967	0,869	0,753	0,601	0,439
−0,80	1,380	1,337	1,285	1,202	1,080	0,953	0,827	0,655	0,473
−1,00	1,525	1,481	1,420	1,317	1,164	1,014	0,878	0,696	0,498
−1,20	1,659	1,610	1,537	1,411	1,228	1,059	0,917	0,725	0,517
−1,40	1,780	1,731	1,639	1,480	1,276	1,096	0,949	0,750	0,531
−1,60	1,897	1,843	1,729	1,533	1,316	1,123	0,973	0,770	0,544
−1,80	2,003	1,947	1,809	1,580	1,347	1,147	0,997	0,787	0,553
−2,00	2,104	2,042	1,879	1,619	1,372	1,167	1,013	0,801	0,560
−2,50	2,340	2,251	2,017	1,690	1,423	1,210	1,049	0,827	
−3,00	2,550	2,414	2,105	1,738	1,457	1,240	1,073	0,840	
−3,50	2,740	2,530	2,153	1,768	1,475	1,252	1,088		
−4,00	2,904	2,609	2,180	1,780	1,487	1,263			
−4,50	3,048	2,671	2,198	1,790	1,491				
−5,00	3,169	2,727	2,207	1,793					
−5,50	3,286	2,769	2,210						
−6,00	3,396	2,800							

Table 26, DSD (1974)

For more details on the subject and example of hydraulic design, it is recommended to consult DSD (1974). The sequence of project activities is described next.

- By trial and error, determine the design head H_o and the crest radius of Figure 3.37 for $\frac{H_0}{R_S} = 0,3$ (recommended).

Note that the flow coefficient of this figure is in English units. For the metric system, multiply by the factor 0,552.

- Determine the discharge curve for a complete range of falls using Figure 3.38.
- Determine the profile of the lower nappe using Figures 3.36 and 3.40 and Tables 3.8 to 3.10.
- Check the throat control in the shaft using equation:

$$R = 0,275 \frac{Q_a^{1/2}}{H_a^{1/4}} \tag{3.9a}$$

where:
C_R = 0,275 for units in the metric system;
R = radius required for throat control (m);
H_a = head of the reservoir for throat (m);
Q = the flow under consideration (m³/s).

If R exceeds the radius of the shaft for a given flow, then the throat control exists, and the flow is based on the radius of the shaft.

- The downstream tunnel is routinely sized to work no more than ¾ filled, with a maximum value of the Manning roughness coefficient $n = 0,016$ to avoid unstable flow conditions. If the tunnel works full, with throat control already developed, the capacity is dictated entirely by full outflow.
- An ideal design has crest control over the entire flow range.

Additional illustrations of morning glory spillway cases are presented here.

Figure 3.44 Classic morning glory spillway. Q = 400 m³/s. Potenza dam/Genzano, irrigation reservoir, H = 84 m, L = 935 m, V = 8,4 Mm³.

Source: (www.pietrangeli.com; Salini Costuttori S.p.A. Bari, Itália).

Figure 3.45 Oued El Makahazine HPP, 36 MW. (1979). Morning glory spillway, $Q = 1.450$ m^3/s, H = 50 m.

Note: Loukkos river, Kenitra, Morocco. Irrigation of the Ksar Plain and Low Hills (3.700 ha).

Dam and reservoir Lower Shing Mun river. Hong Kong (*wsd.gov.hk*)

(*wsd.gov.hk*)   (greenpower.org.hk)

Figure 3.46 Reservoir on the Lower Shing Mun river, Hong Kong. Morning glory spillway, $Q = 425$ m^3/s.

Note: In the morning glory region, excavations were made to improve the approach flow and, consequently, the performance of the structure.

3.6.6 Side spillway design example

As stated in section 2.3, a side spillway can be used when the topographic characteristics of the site imply a solution for the arrangement of the dam works with the spillway not incorporated into the axis of the main dam of the river.

For example, if the project flow is $Q = 1.000$ m^3/s and the available width for spillway is $L = 100$ m, assuming $C = 2,1$, at the crest we will have $Ho = (Q/L/C)^{2/3} = 2,85$ m.

A practical example of hydraulic design of the waterline profile along the side channel trough is presented in the USBR DSD (1974), section 202, including the spreadsheet (see Figure 2.2).

Follow the illustrations of the classical example of the two side spillways of Hoover dam (1936), Colorado river, Nevada/Arizona (USA), between Las Vegas and Boulder City. This dam is 180 m high and each spillway has four gates ($L = 30,48$ m; $H = 4,88$ m). The total discharge capacity is of the order of 11.330 m^3/s.

A

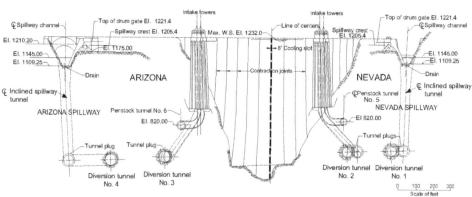

Figure 3.47 Hoover dam/Lake Mead (1936).

Source: Colorado river, Nevada/Arizona (USA).

Chapter 4

Hydrodynamics pressures

This chapter presents the hydrodynamics pressures caused by the flow on the spillway. Figure 4.1 shows photographs and flow nets of the flow above the Creager standard profile crest, the latter having a prominent nose (see also section 3.2).

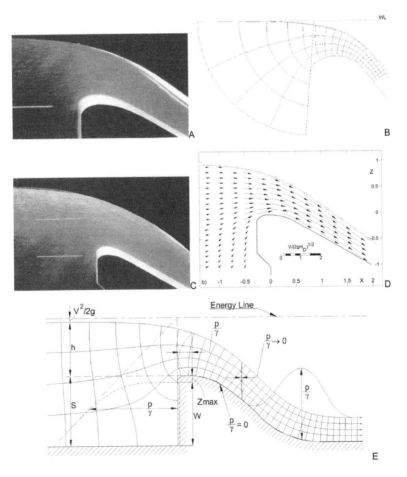

Figure 4.1 Flow over the spillway. A and C: Standard crest, Creager type profile; B, D and E: Flow nets.

Source: A, C and D: Vischer and Hager (1998); B: Cedergren (1967). E: Rouse (1938, 1950).

4.1 Initial considerations

For an ideal profile, designed exactly with the shape of the inferior nappe over a thin sill, the pressure on the crest for the design head would be zero (atmospheric pressure).

For practical reasons, however, such an ideal profile is generally modified in such a way that, under the design head, low pressures will develop.

The spillway will operate at different heads of the design discharge: for low heads, the pressure will increase, and for high heads, the pressure will decrease.

Assuming that the two-dimensional flow is irrotational, the pressure at the crest of the spillway can be precisely determined:

- analytically by a numerical method;
- instrumentally by an electronic analogy (see Chow, 1959);
- graphically using a flow network;
- in tests on reduced two-dimensional models (see section 4.5 for the measurements made in the Tucuruí model).

Users are advised to refer to the work of Dao-Yang and Man-Ling, presented at the 13th ICOLD (New Delhi, 1979): "Mathematical Model of Flow Over a Spillway Dam."

Flow over the structure involves curved stream lines, with origin of the curvature in the bottom stream (Figure 4.1B). The gravitational component of a fluid element is then reduced by centrifugal force. If the curvature is large enough, the internal pressure may fall below atmospheric pressure and may reach values below the vapor pressure, with risk of cavitation damage, which is unacceptable.

For optimum performance of the spillway, as already stated, the flow must be well accommodated, without detachment and disturbance of the liquid vein, for the various operating conditions of the structure. Experience has shown that an ideal situation is not always achieved.

In the crest region, there is a significant increase in the velocity of the bottom current and the current in the flow direction, disturbing the pressure field. Downstream of the crest, the velocity distribution is uniform, and the pressure distribution is almost hydrostatic.

The distribution of these mean pressures (p/γ) along the spillway profile, with and without piers, are shown in the dimensionless USBR graphs (see Figures 4.2, 4.2A and 4.2B).

Down the profile, the flow accelerates and implies hydrodynamics pressures on the structure, which are given by the sum of the mean pressures (p/γ), with the instantaneous pressure fluctuations ($v^2/2g$).

Fluctuations are as large as they are at flow velocities and can cause cavitation damage to the structure (see Chapter 8), as well as scour downstream of the structure (see Chapters 5 to 7).

4.2 Average pressures

The mean pressures (hp) can be estimated:

- using the graphs of Figures 4.2, 4.2A and 4.2B (HDC 111–16 to 111–16 / 2) for the case of flow without control of gates;
- using the data of Figure 4.3 (HDC 311/6, or Lemos – 1965) for the case of flow with control of gates.

In the case of ski jump spillways, bucket pressures can be estimated using the graph of Figure 4.4 (HDC 112–7).

This graph represents plots of mean pressure data measured in a model for a standard crest with and without piers. Additionally, it is recommended to consult Balloffet (1961). The data are normalized: $hp/H_d \times X/H_d$, for several values of H/H_d, where:

hp = pressure (m);
H_d = project load (m);
X = the abscissa, measured from the crest of the spillway, as shown in Figure 4.2.

The pressures for intermediate head relations can be obtained by interpolation.

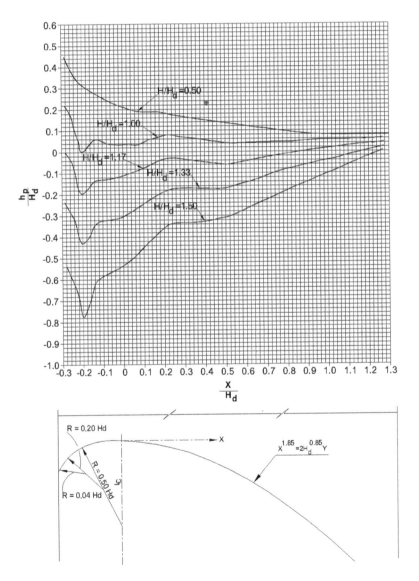

Figure 4.2 Crest pressures – without piers.

Source: HDC chart III–16 (1959).

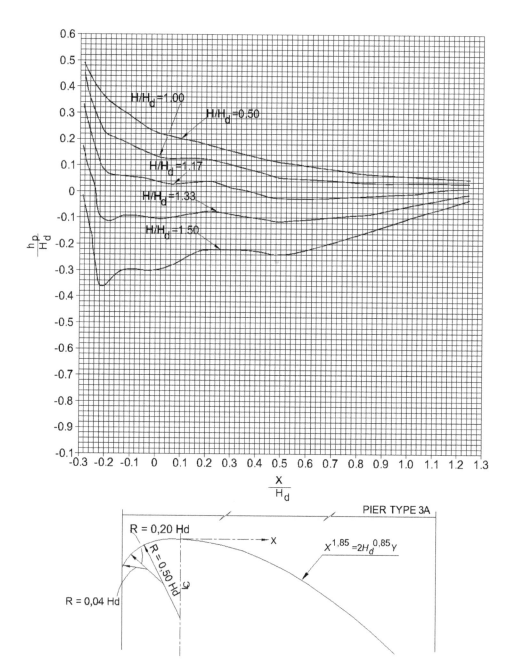

Figure 4.2A Crest pressures: in the center line of the gap between the piers.
Source: HDC chart 111–16/1 (HDC, 1959).

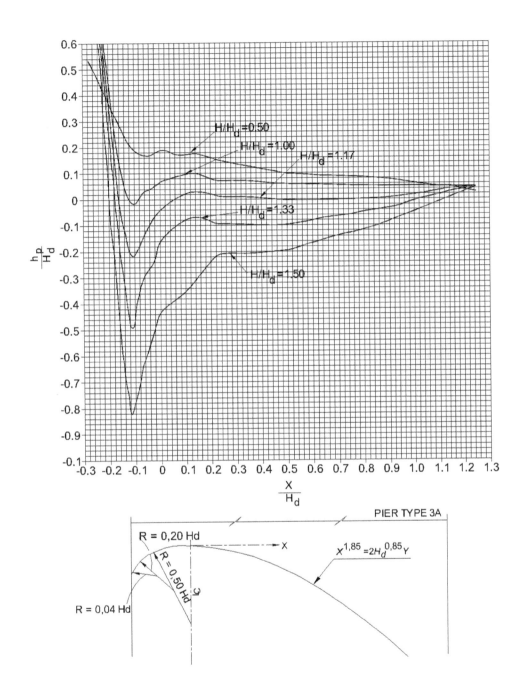

Figure 4.2B Crest pressures: along the piers.
Source: HDC chart 111-16/2.

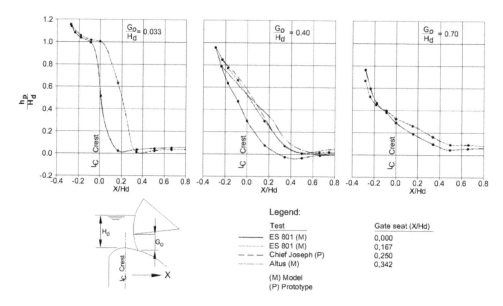

Figure 4.3 Crest pressures: gated spillway.
Source: HDC chart 311–6.

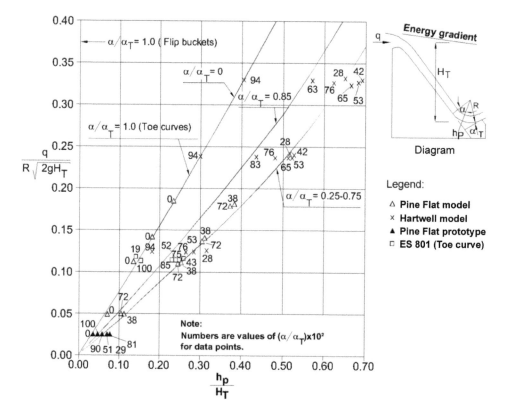

Figure 4.4 Pressures in the bucket and in the flip bucket.
Source: HDC chart 112–7.

4.3 Instant pressures – Shin-Nariwa (Japan)

The instantaneous pressures fluctuations, caused by the high velocity flow on the rough surface of the concrete in the turbulent boundary layer, are measured in model and proto- type. These fluctuations can also be estimated using the Kraichnan equation (1956).

The case of the Shin-Nariwa dam (Figure 4.5), built on the Nariwa river by the Chugoku Electric Power Company, –Japan, presented in detail by Minami (1970), is used as an example.

This dam, 103 m high, has a ski jump spillway with capacity of 2.400 m³/s. The slab of the basin before the ski jump, 50 m long and 60 m wide, is the roof of the powerhouse.

The reservoir water level is at elevation 237,00 m and the slab at elevation 161,00 m. The difference is 78 meters. Therefore, the flow on the chute/slab has a velocity greater than 30 m/s.

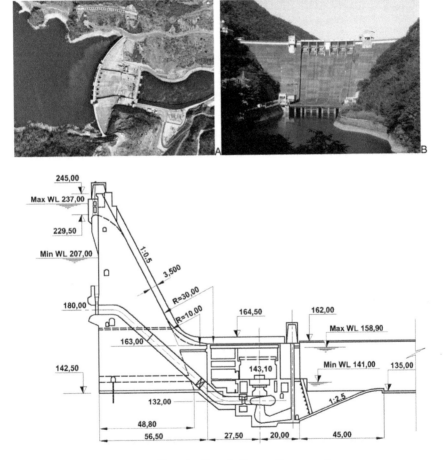

Figure 4.5C Shin-Nariwa dam. Chugoku Electric Power Company, Japan.
Source: (Minami, 1970).

Minami (1970) measured the velocities and pressures distribution in the model and pro- totype of the Shin-Nariwa Power Plant. The pressure cells were installed as shown in Figure 4.6. The results are presented in Figures 4.7, 4.8 and 4.12.

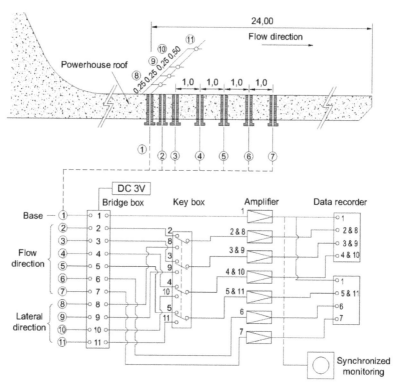

Figure 4.6 Shin-Nariwa dam. Arrangement of cells pressure.

Source: (Minami, 1970).

It should be noted that the flow is divided in the turbulent boundary layer, where there is velocity gradient, and in the potential flow, where the velocity distribution is uniform and varies according to the flow.

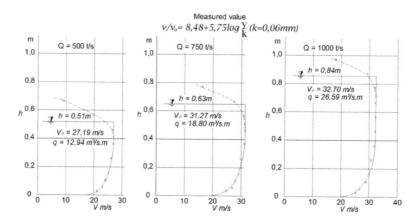

Figure 4.7 Distribution of velocities on powerhouse roof slab – prototype results.

Source: (Minami, 1970).

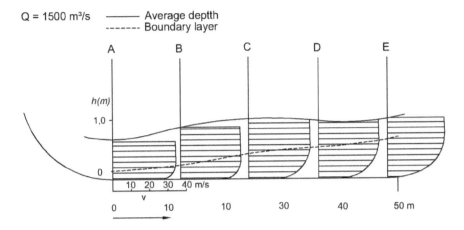

Figure 4.8 Distribution of velocities on powerhouse roof slab – model results.

Source: Minami (1970).

The characteristics of the flow velocity distribution in the turbulent boundary layer in the slab model are well represented by the von Kármán/Nikuradse equations for flow in smooth tubes, as shown in Figure 4.9 (Rouse, 1938, republished in 2011).

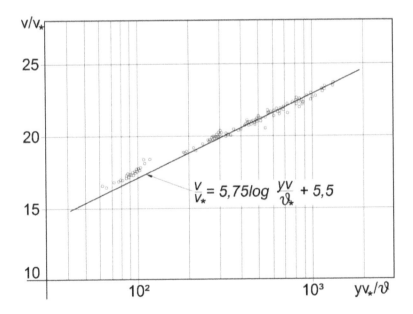

$$\frac{v}{v_*} = 5{,}75 log \frac{yv}{\vartheta_*} + 5{,}5$$

Figure 4.9 Profile of velocities on the slab in the model.

Source: Minami (1970).

Minami (1970) analyzed the characteristics of velocity distribution in experiments in the prototype. Unlike the model, he assumed that the prototype would be worth von Kármán/Nikuradse's equations for rough pipes.

Considering the concrete surface with a K roughness, the velocity distribution can be represented by the following equations:

$$\frac{v}{v^*} = 8,48 + 5,75 log \frac{y}{K} \qquad (4.1)$$

$$\frac{v_o}{v^*} = 8,48 + 5,75 log \frac{\delta}{K} \qquad (4.2)$$

where:

$v^* = \sqrt{\tau_o / \rho}$ = friction velocity;

τ_o = shear stress in concrete (t/m^2);

ρ = density of water (t s^2 m^{-4});

v = velocity in the boundary layer at a point distant "y" from the surface (m/s);

v_o = flow velocity (m/s);

δ = thickness of boundary layer (m);

K = concrete roughness (0,0006 m).

The shear stress in the concrete, using the coefficient of friction, is represented by the following equation.

$$\tau_o = \rho v_o^2 = C_f \frac{\rho v_o^2}{2} \qquad (4.3)$$

From equations (4.2) and (4.3) we have:

$$C_f = \frac{2}{\left(8,48 + 5,75 log \frac{\delta}{K}\right)^2} \qquad (4.4)$$

From equations (4.1) and (4.3) we have:

$$\frac{v}{v_o} = 8,48 \sqrt{\frac{C_f}{2}} + 5,75 \sqrt{\frac{C_f}{2}} \, log \frac{y}{K} \qquad (4.5)$$

Equation 4.5 provides the velocity distribution in the boundary layer as a function of C_f and y/K. Then, using the experimental value v/v_o and y, a combination of C_f and K that represents the best relation of equation (4.5) can be determined graphically.

As shown in Figure 4.7, the combination of $Cf = 0,0022$ and $K = 0,06$ mm represents the best ratio of experimental values. Then the thickness of the boundary layer can be calculated from equation (4.4): $\delta = 0,35$ m.

It should be noted that δ can also be calculated using the charts HDC 111–18 to 111–18/5:

$$\delta/L = 0,08(L/K)^{-0,233} \qquad (4.5a)$$

where

L (m) is the length of the crest of the spillway – chart HDC 111–18/1;

K is the roughness of the concrete.

It should be noted, furthermore, that these charts are entitled "Energy Losses in Spillways." Energy is lost in the boundary layer.

The estimation of these losses may be important for the design of the energy dissipation device downstream of the spillway, either for the stilling basin or for the jet throw distance from a ski jump.

The curves shown in Figure 4.10 (of the model) are the result of the calculations of Equation 4.5) and prove that the measured values of v and δ coincide with their calculated values.

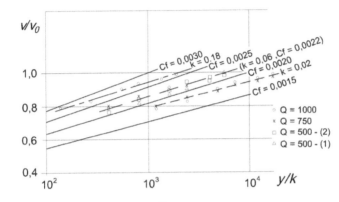

Figure 4.10 Relations between v/v_o and y/K

Source: Minami (1970).

The values measured in prototype arranged in the form of Equation 4.1 are shown in Figure 4.11.

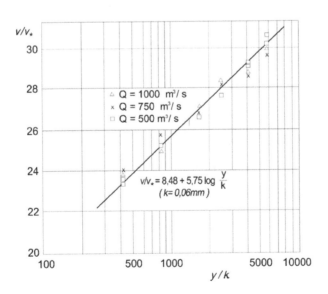

Figure 4.11 Profile of velocity on the slab in prototype.

Source: (Minami, 1970).

Comparing these values with those of the model, it can be seen that the velocity distributions are in agreement with the von Kármán/Nikuradse equations.

Forces acting in the chute/powerhouse roof slab

The forces acting on the chute are of a random oscillatory nature due to the turbulent characteristics of the high velocity flow, which are divided into average values and floating values. The mean values of the vertical and tangential components are due to the static pressure due to the water rod and the shearing stress due to the friction. These forces can be calculated by Equations (4.3) and (4.4).

The pressure fluctuations in the turbulent boundary layer, of a random nature, were measured in the prototype. The maximum flotation was of the order of 3,0 t/m^2 and the frequency was in the order of 5 to 300 c/s. For more details see Minami (1970).

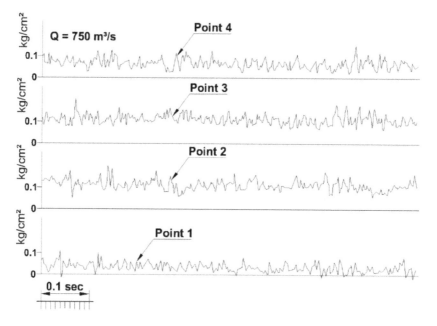

Figure 4.12 Pressure fluctuations in the slab in the prototype.
Source: (Minami, 1970).

Pressure fluctuations were also estimated using the Kraichnan equation:

$$\sqrt{p_w^2} = \beta C_f \frac{\gamma V_E^2}{2g} \tag{4.6}$$

where:

$\sqrt{p_w^2}$ = standard deviation of pressure fluctuations (m);
β = coefficient that normally ranges between 3 and 5;
C_f = coefficient of friction, calculated at each point according to equation (4.4);

βC_f = the coefficient of pressure fluctuation $\acute{C}_p$;
V_E = velocity of the flow (m/s); $V_E = q/d$ (specific discharge/height of water); can also be calculated by the boundary layer method;
γ = specific weight of water = 1,0 t/m^3;
g = acceleration of gravity = 9,81 m/s^2.

Studies by Serafini correlated β with the boundary layer, as shown in Figure 4.13. Defined β, we have all the elements to estimate the fluctuations of instantaneous pressures.

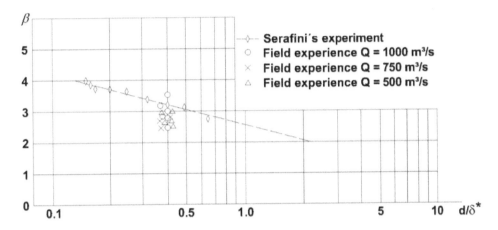

Figure 4.13 Relation between β and δ. d = diameter of the pressure cell; $\delta^* = \delta_1 = 0{,}18\ \delta$ (HDC 111–18).

Source: (Serafini, 1963)

Powerhouse roof slab oscillations

The maximum pressure amplitude in the powerhouse roof was 3,0 t/m^2 for the discharge of 1.000 m^3/s. But because of the large frequency range and the limited phase simultaneity scale, the pressure fluctuations were not as severe as the vibration forces against the powerhouse roof slab and did not give any unpleasant oscillations. The analyses of tensions and oscillations in the roof of the powerhouse showed that the natural frequency of the roof slab was of the order of 10 c/s and the maximum displacement value of the order of 10 μ for the flow of 1.000 m^3/s. These results were published by Suzuki (1973). The summary presented here aims to compose the book and show a way for the young engineers to begin the study of this complex matter, which has been much explored in the field of aerodynamics. It is recommended to consult the cited references.

Variation of $\acute{C}$p

It should be noted that in the specialized literature, the product βC_f of Equation 4.6 is the pressure fluctuation coefficient $\acute{C}p$. Pinto, L. C. S. et al. (1988 and 1991) presented a work on the macro turbulence of the hydraulic summit in dissipation basins, from where we extract the data presented in Table 4.1 and in Figure 4.14.

Table 4.1 Values of $\acute{C}p$

Authors	FI	$\acute{C}p$ max	X/Y_1
Khader and Elango	4,7	0,080	$8 \leq X/Y_1 \leq 12$
	5,9	0,082	
	6,6	0,085	
Camargo	5,1	0,069	$X/Y_1 = 12,5$
Lopardo and Henning	4,66	0,065	$8 \leq X/Y_1 \leq 12$
	6,30	0,069	
	9,54	0,050	
Pinto and Vasconcellos	7,97	0,051	$7,9 \leq X/Y_1 \leq 16$
	9,11	0,043	
	10,06	0,039	

Source: (Pinto and Vasconcellos, 1988).

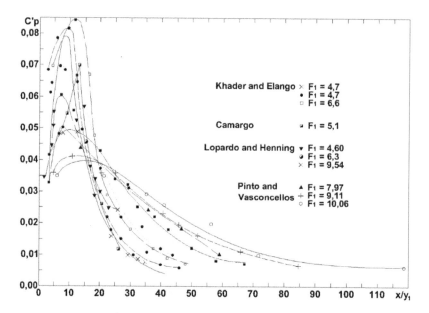

Figure 4.14 Variation of $\acute{C}p \times X/Y_1$. X = abscissa of the pressure taking; Y_1 = upstream depth of the hydraulic jump.

Source: (Pinto and Vasconcellos 1988)

Figure 4.15 shows the same type of graph extracted from Marques (1995) of the IPH-UFRGS, also for the downstream jump of spillways.

Figure 4.16 shows the results of Lopardo *et al* (1987) for jumps with asymmetric maneuvers of gates. He showed that asymmetric spillway flooding operations increase turbulence and pressure fluctuations in the stilling basin, as can be seen in the following graph.

Other examples of projects similar to Shin-Nariwa:

- the Karakaya dam (Stutz, 1979), presented in the next section;
- the Ivailograd dam on the Arda river and the Antonivanovtsi dam on the Vacha river, both in Bulgaria (Stanchev, 1973);

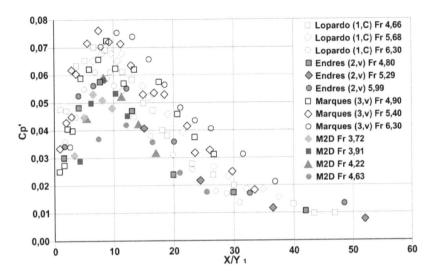

Figure 4.15 Variation of $\acute{C}p * \times X/Y_1$. X = abscissa of the pressure taking; Y_1 = upstream depth of the hydraulic jump.

Source: (Marques, 1995)

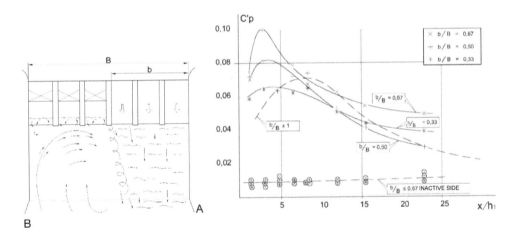

Figure 4.16 Schematic plant of a spillway and variation chart of $\acute{C}p$ in asymmetric maneuvers.

Source: (Lopardo 1987).

- WP & DC (April 1989) presents the case of Fengtan HPP, 800 MW (Figure 2.1C and Figure 5.38), on the You river, Zhangjiabe, Hunan Province, China; it is an arched dam with $H = 112,50$ m and $L = 488$ m; spillway with radial gates for 23.300 m³/s;
- Soos (1982) proposed such a solution for the Xingó HPP;
- Sharma (1982) in his paper shows a project proposal in India but, for some reason, does not cite the name.

4.4   Instant pressures – Karakaya

The Karakaya power plant (1.800 MW) (see Figure 4.17), near the city of the same name, is located in a narrow valley of the Euphrates river in Turkey, between Atatürk (2.100 MW) and Keban (1.240 MW). The cross-section of the dam incorporates the powerhouse and the spillway into a single block (Figure 4.18). The dam is 180 m high and 392 m long – crest length. The spillway in ski jump (Figure 4.19), with ten segment gates (H = 14,43 m × L = 14 m) and has a capacity of 22.000 m³/s. The maximum specific discharge is 130 m³/s/m (Idrotecnica, 1988).

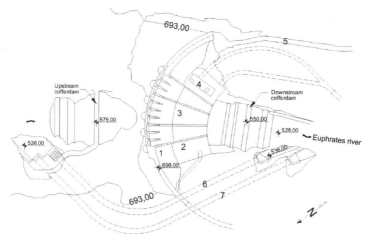

1 - Dam 2 - Powerhouse 3 - Spillway 4 - Service building 5 - Access road 6 - Diversion tunnel, bottom outlet after construction 7 - Diversion tunnel, closed after construction

Figure 4.17 Arrangement of the Karakaya power plant.
Source: (Stutz, 1979).

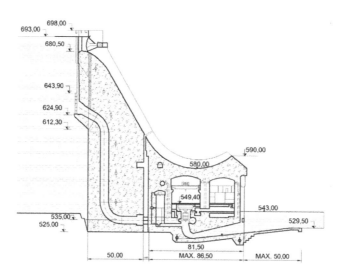

Figure 4.18 Karakaya HPP. Dam, spillway and powerhouse.
Source: (Stutz, 1979).

Figure 4.19 Karakaya HPP. Ski jump spillway working.

Source: (Idrotecnica, 1988).

The mean pressures due to gravity force and centrifugal force, shown in Figure 4.20, were measured and calculated in the classical manner, as previously reported.

$$\frac{P_m}{\gamma} = H - (H - d\cos\emptyset)\left(\frac{R-d}{R}\right)^2 \tag{4.7}$$

where:
H = head on the bucket (m);
d = depth of water on the bucket (m);
R = bucket radius (m);
ϕ = angle between the radius tangent and the horizontal in each point along the bucket (m);
γ = specific weight of water (t/m^3).

In the transition from the bucket to the straight stretch $R = \infty$ and Equation 4.7 becomes $Pm/\gamma = d$, which is the hydrostatic pressure.

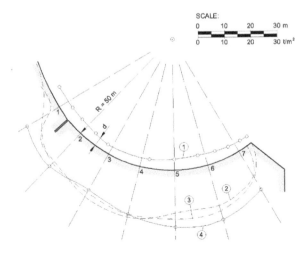

Figure 4.20 Pressures on the powerhouse roof (in spillway axis). (1) Measured waterline depths. (2) Measured water pressures. (3) Adjusted pressures. (4) Theoretical pressures.

Source: (Stutz, 1979).

The pressure fluctuations, shown in Figure 4.21, were calculated using the Kraichnan equation using $C_f = 0,013$.

According to Stutz (1979), because of the small importance of these forces, no vibrations of the structure were expected because of the randomness of the amplitudes and wide frequency range outside the resonance frequency range of the powerhouse structure. This was confirmed by the results of application of the Shin-Nariwa HPP (previous case), where the calculations corresponded to both the model and prototype results.

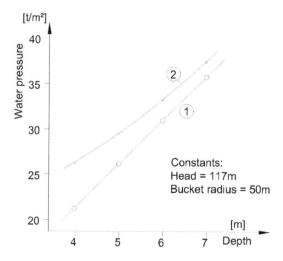

Figure 4.21 Pressures × water depth. (1) Pressures due to gravity force and centrifugal force. (2) Pressures + fluctuations.

Source: (Stutz, 1979).

It should be noted that in the joint between the dam and the powerhouse an aerator was provided, as shown in Figure 4.22. Air demand was estimated at 8% of the maximum specific discharge of 130 m³/s/m.

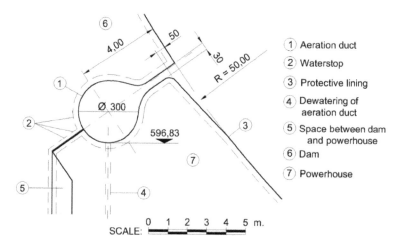

Figure 4.22 Aerator: joint between the dam and powerhouse.

Source: (Stutz, 1979).

4.5   Instant pressures – Tucuruí (Brazil)

The main characteristics of the Tucuruí spillway are:

- 23 spans of 20 m width at the crest (Figure 4.23);
- width on the crest = 460,00 m; width on the jet apron lip = 552,00 m;
- length of the spillway profile = 90,00 m.
- head on the jet apron lip 44 m;
- maximum capacity of discharge of 110.000 m^3/s; maximum specific discharge = 228,00 $m^3/s/m$;
- velocity in the bucket = 31 m/s; velocity on the jet apron lip = 28,40 m/s.

Figure 4.23 Finishing the concrete surface (1984).

Figure 4.24 Inspection after the first year of operation, in October (1985), by engineer Selmo Kuperman and the author near the jet apron lip.

Due to the size of the structure, a two-dimensional model was available in the 1:50 scale for several studies and surveys, including hydrodynamic pressure surveys on the structure for various operational situations. These studies are presented in Chapter 10.

The mean pressures were recorded in several pressure gauges shown in Figure 4.25. The hydrodynamic pressure were recorded only in the gauges 9, 25 and 25A. The enveloping piezometric lines are also shown in Figure 4.25.

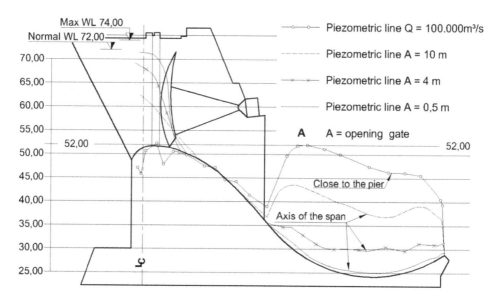

Figure 4.25 Location of pressure gauges and enveloping piezometric lines.
Source: (Magela, 1983).

The mean pressures and the instantaneous fluctuations recorded in the hydraulic model were close to those theoretically estimated: the mean pressures by the methods of Balloffet *et al.* (1948) and Lemos (1965); the fluctuations through Kraichnan's (1956) equation, quoted by Minami (1970).

It was concluded that there is no risk of cavitation in the structure. However, during the project, a detailed survey of the incipient cavitation index was carried out in 1983 based on Ball (1976) and Pinto (1979). The results are presented in Chapter 8.

It should be noted that the technical specifications for finishing the concrete surfaces of the spillway limit gradient irregularities by 1:50, with $d = 0,60$ cm.

Using Figure 8.2, $Ki = 0,10$ (critical) can be obtained for this type of irregularity, but with $d = 0,70$ cm. The value found by Magela (2015a) for the end of the Tucuruí spillway was $Ki = 0.40$, for flows of 100,000 and 110,000 m^3/s. Considering that $d = 0.60$ cm in Tucuruí, security increases.

In spite of this, a rigorous inspection of the finishing of the concrete surfaces of the spillway was carried out in 1984 (see Figure 4.24). At the time, it was verified that the finish was still very good.

This spillway has been operating with some frequency since October 1984, and its performance can be considered very good, as reported by Magela *et al.* (2015). However, some small problems of cavitation erosion were recorded along the slots of the stoplog of the gates (see Figure 5.1B).

Chapter 5

Energy dissipation

5.1 Classification of the dissipators

In general, energy dissipators are classified according to the following basic concepts (see Figures 5.1A to 5.1F:

- those in which the dissipation occurs through the hydraulic jump and the stilling basins, for small to medium heads up to 30 m;
- those in which dissipation occurs through the ski jump, or dispersion of the water jet in air, for heads larger than 50 m;
- those intermediate to hydraulic jump and a ski jump called plunge pools of drop structures;
- those in which dissipation occurs in a slope chute, for heads larger than 50 m;
- those in which dissipation occurs in steps.

The criteria for defining dissipators are loosely defined and depend on the experience of the designer. In this task, the following factors are involved:

- topography, geology of the dam site and type of dam;
- the layout of the plant (general arrangement of works);
- hydraulic parameters: head, discharge and flow rating curve; and
- economic comparisons between different types of dissipators.

Figure 5.1A Stilling basin.
Source: Porto Colômbia HPP, Brazil.

Figure 5.1B Ski jump. Tucuruí HPP, Brazil.

Figure 5.1C Ski jump. Tarbela HPP, Pakistan.

Figure 5.1D Plunge pool of drop structure. Katse, Lesotho.
Source: Katse dam, Lesotho.

Figure 5.1E Plunge pool of drop structure. Wagendrift dam, South Africa.
Source: Vischer and Hager (1998).

Figure 5.1F La Grande 2 spillway – Steps (Hydro Québec).

In addition to these, other important factors must be considered: frequency of operation and facilities for maintenance of the spillway; associated risks of damages and ruptures; and the experience of the designer.

Given the complexity of the issue, even observing all these factors, erosions have been recorded downstream of the spillways. What is important is to prevent them from threatening the stability of the works like the dam and the spill itself.

It should be noted that the reduced hydraulic models, used as an additional tool in the study of dissipators, have limitations, since they cannot reproduce some aspects that affect the prototypes, such as the aeration degree of the flow and the resistance of the rock mass.

It is observed that the solution of the ski jump provides a great economy in civil works, when compared to the solution of a stilling basin, and this is an argument to consider in favor of this type of dissipator when there is sufficient hydraulic head for the jet takeoff. In the case of Tucuruí, it meant a saving of US$ 100 million, at 1988 values.

It would not be technically and economically acceptable to think of adopting a concrete dissipation basin to protect the rocky massif downstream that did not have the best geomechanical conditions to withstand, without erosion, the hydrodynamic flow pressures.

To overcome this, a downstream pre-excavation was designed to function as a plunge pool and as a quarry for the rockyfill dam of the central channel of the river. This spillway has operated very frequently since November 1984, according to plan, without significant erosions.

This theme is dealt with in several references on the references list, of which the following can be highlighted:

*   Berryhill, R. H. "Stilling Basin Experiences of the Corps of Engineers." *Journal of the Hydraulics Division*, ASCE, Vol. 83, HY 3, June (1957);
*   Berryhill, R. H. "Experience with Prototype Energy Dissipators." *Journal of the Hydraulics Division*, ASCE, Vol. 89, HY 3, May (1963). Several cases are presented in this paper;
*   Berryhill, R. H. et al. "Energy Dissipators for Spillways and Outlet Works. Progress Report of the Task Force on Energy Dissipators for Spillways and Outlet Works." Committee on Hydraulics Structures. *Journal of the Hydraulics Division*, ASCE, Vol. 90, HY 1, Jan (1964).
*   Question 41 of the 11th ICOLD, "Control of Flows and Energy Dissipation during Construction and after Construction," Madrid, 1973;
*   Question 50 of the 13th ICOLD, "Large Capacity Spillways and Orifices," New Delhi, 1979;
*   Report on the "Spillway Performance Energy Dissipation, Cavitation and Erosion," XIII SNGB, Rio de Janeiro (Brazil), April 1980, by Neidert, S.H. Reference: CEHPAR Publication 37, Curitiba, Brazil (1980);
*   *Design of Energy Dissipators for Large Capacity Spillways*, Rao, K. N. S. The Transaction of the International Symposium on Layout of Dams in Narrow Gorges, Rio de Janeiro (1982).
*   *Advanced Dam Engineering for Design, Construction and Rehabilitation*, by Jansen, R. B. (1988).

As mentioned by Neidert (1980b), every dam is associated with a head, and in its transposition flow flows and the consequent acceleration of flow. The energy dissipation in the transposition is small, because, for practical and economic reasons, the flow path is usually short and is made on smooth contours.

Exceptions to this general rule are so-called high-roughness chutes in works where topographic and geological conditions are favorable, as in the case of the La Grande 2 spillway, Figure 2.40 (Aubin, 1979), and Serra da Mesa spillway, Figures 2.25 and 2.26 (MDB, 2000).

High concentrations of energy are therefore normal at the base of the spillways, and important dynamic actions are common in these regions, leading to negative reflexes of various natures. Two basic parameters define the value of the energy:

*   the velocity (V) at the outlet of the structure or the impact position in the riverbed downstream – implicitly the difference (H), less energy losses; and
*   the flow concentration (q).

Its quantification can be made from the quantity of movement (QM) and the power to dissipate per meter of width of the structure (N), among other forms as shown Equations 5.1

and 5.2, disregarding $Cp = 1,0$ in favor of safety – see Aubin (1979) and Neidert (1980a,b).

$$QM = 101,9 \ q \ V \ = 101,9 \ q \ \sqrt{2 \ g \ H} = 451,5 \ q \ \sqrt{H}(t/m) \tag{5.1}$$

$$N = 9,81 \times 10^{-3} \ q \ \frac{V^2}{2g} = 0,01 \ q \ H(MW/m) \tag{5.2}$$

Neidert (1980a,b) presented these parameters for several Brazilian spillways: Table 5.1 and graphs of Figures 5.1 and 5.2.

Table 5.1 Brazilian spillways, examples (Neidert, 1980a,b).

HPP	Discharge (m³/s)	H (m)	Q (m³/s/m)	N (MW/m)	QM (t/m)
1-Furnas	13.000	85,23	153,00	128,00	637,70
2-Estreito	13.000	51,09	155,50	77,90	501,80
3-P. Colômbia	16.000	32,70	98,20	31,50	253,50
4-Marimbondo	21.400	76,56	131,30	98,60	518,70
5-Itumbiara	16.200	72,25	162,00	114,80	621,70
6-Água Vermelha	20.000	43,30	135,10	57,40	401,40
7-Três Irmãos	9.500	42,50	131,90	55,00	388,20
8-Ilha Solteira	40.000	68,50	140,40	94,30	524,30
9-Xavantes	3.200	89,00	66,70	58,20	284,10
10-São Simão	21.400	65,80	154,50	99,70	565,80
11-Volta Grande	12.700	39,70	72,00	28,00	204,80
12-Jaguara	14.000	32,50	172,80	55,10	444,80
13-Três Marias	8.700	51,50	108,80	55,00	352,50
14-Emborcação	7.800	83,50	132,90	108,90	548,30
15-S. Santiago	24.100	34,84	160,20	54,80	426,90
16-S. Osório I	15.000	59,00	158,70	91,90	550,40
17-S. Osório II	12.500	45,30	170,70	75,90	518,70
18-Foz do Areia	11.000	120,50	155,80	184,20	772,20
19-Coaracy Nunes	11.000	23,80	62,00	14,50	136,60
20-Tucuruí	100.000	44,00	180,00	77,70	539,10
21-Itaipu	62.000	85,00	185,00	154,30	770,10

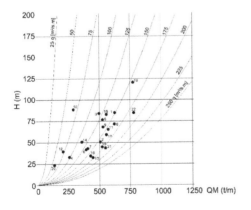

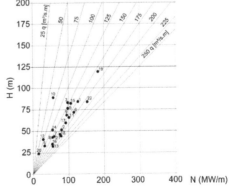

Figure 5.2a Spillways: quantity of movement per meter wide.

Figure 5.2b Spillways: power per meter wide.

Source: (Neidert, 1980a,b).

Source: (Neidert, 1980a,b).

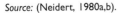

5.2   Ski jump dissipators

5.2.1   *Summary of theory*

In a ski jump, the kinetic energy of the flow is used for the takeoff of the jet in the air (Figure 5.3A). The energy is mainly dissipated by the jet dispersion (jet disintegration). The throw distance is calculated using the HDC chart 112–8 – Figure 5.3.

The energy is dissipated mainly in three phases:

- in the aerial phase of the jet by the air resistance, which reduces its throw distance;
- in the submerged phase when there is a tailwater depth, in the process of diffusion of the jet, attenuating its erosive power; and,
- in the rocky massif, after impact.

The jet path changes as the throw distance decreases, the cross-section area increases (relative to the theoretical value) and the jet energy decreases. Data referring to this evolution and the estimation of effective reach are presented by Martins (1977).

The process of jet diffusion in the downstream water mass can be accurately analyzed, according to Hartung and Häusler (1973), using the theory of turbulence of free jets, as will be seen in Chapter 6.

The energy dissipation occurs in the process of momentum exchange: the jet at velocity penetrates the body of water downstream at rest. In this case, the dispersion is almost completely linear.

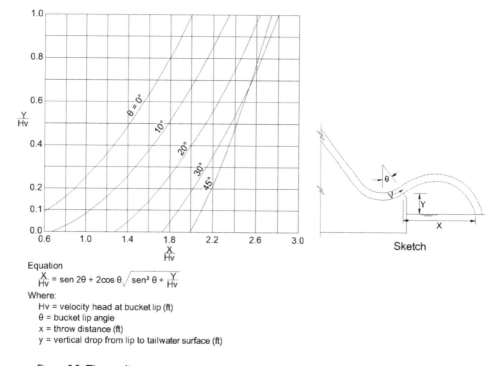

Equation

$$\frac{X}{Hv} = \text{sen } 2\theta + 2\cos\theta \sqrt{\text{sen}^2\,\theta + \frac{Y}{Hv}}$$

Where:
Hv = velocity head at bucket lip (ft)
θ = bucket lip angle
x = throw distance (ft)
y = vertical drop from lip to tailwater surface (ft)

Figure 5.3 Throw distance.

Source: Chart HDC 112–8 (HDC, 1959).

Figure 5.3A Tucuruí spillway. Jet throw distance (Eletronorte).

After these two phases, aerial and submerged, the residual energy of the jet is transmitted to the rocky massif, with the consequent development of the scour pit.

The development of the scour pit was presented in detail by Spurr (1985). See also Sobrinho and Infanti (1986). Research on the development of the erosion pit is done in qualitative trials on hydraulic models. The estimation of the erosion pit is presented in Chapter 7.

The impact of the jet exerts a combination of forces on the rocky massif, which vary as the scour develops. Initially, when the pit (or pre-excavation) is shallow, the pressure gradient is high, and the rock mass will deflect the jet.

The penetration of high pressures in the stratification planes, faults and fractures causes the hydrofracturation of the massif, and scour develops. The pressure propagation rate is a function of the fracture degree, the fracture conditions (open, sealed, decomposed) and the orientation of the discontinuities in relation to the angle of incidence of the jet.

When the rock mass is homogeneous and strong enough to confine the currents in the jet impact zone, the hydraulic action on the contours of the pit will be more intense. In this case, erosion is concentrated in this region and deepens, as happened in Jaguara HPP, where the jet impacted, at the beginning, directly on the rocky massif, without protection of the tailwater depth. Erosion developed rapidly in the first troughs. This case was presented in detail by Magela and Brito – available on the internet (1991).

5.2.2 *Spillway of Jaguara*

As it is a remarkable case, it was considered opportune to represent some photos of the erosion obtained in trip to the plant in July 1989, showing details of the rocky massif, composed of a sound quartzite, very fractured, with no decomposed zone (Figures 5.4 to 5.10. The spillway in Jaguara is positioned on the same axis of the dam, as can be seen in Figures 5.4 and 5.8.

In 2001 the owner built a channel downstream of the spillway shown in Figure 5.9 (CBDB, 2002).

Figure 5.4 Jaguara HPP. Spillway: Q = 12.600 m³/s.
Source: (Magela and Brito, 1991).

Figure 5.5 Downstream slope of the pit. Aspects of the rocky massif: sound quartzite, fractured.

Figure 5.6 Details of fractures and joints.

Figure 5.6A Fractures and joints.

Figure 5.6B Jaguara HPP. Details of the fractures and joints of the rocky massif downstream. X = schistosity; JD = distension joint; JC = shear joint.
Source: Magela and Brito (1991).

Figure 5.7 Jaguara HPP. Downstream view: Spillway, opening 01. Note: Dam protection wall near the slope of the deep natural channel of the right bank

Source: (Magela and Brito, 1991).

Figure 5.8 Jaguara HPP. Scour pit, 1982.

Source: (CBDB, 1982a,b).

Figure 5.9 Jaguara HPP. New channel downstream of the spillway.
Source: (CBDB, 2011).

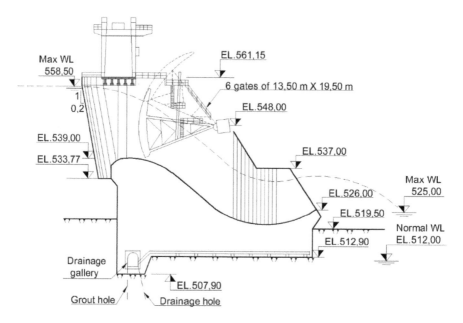

Figure 5.10 Jaguara HPP ski jump spillway.
Source: (CBDB, 1982a,b).

5.2.3   Spillways of Tarbela

In heterogeneous masses and where there is low resistance, lateral scour can rapidly develop asymmetrically and evolve toward the walls or the foot of the dam, as happened downstream of the service spillway of the Tarbela dam. The dam was implemented between 1968 and 1974 on the upper reaches of the Indus river, Pakistan, near the city of Islamabad, with the purpose of regulating flow, irrigation and power generation – 6.500 MW (Lowe, 1973) and Haüsler (CBDB/ICOLD, 1982b).

The plant has a rockfill dam 2.740 m long and 143 m high and is founded in thick alluvium. The spillways have a total discharge capacity of 42.470 m^3/s. The auxiliary spillway has nine gates and capacity of 24.070 m^3/s. The service spillway is 106,70 m wide, with seven gates and a discharge capacity of 18.400 m^3/s. The difference between the level of the reservoir and the bucket lip is 100 m, which implies very high jet velocities of more than 40 m/s. The difference between the level of the bucket lip for the downstream rock in the plunge pool was only 6,0 m. The elevation of the rock was 365,80 m, and the minimum downstream WL was 347,50 m. For the design discharge, the downstream WL rose to elevation 363,00 m. That is, the jet impacted directly on the rock.

The preponderant geological feature in the basin was an igneous intrusion that cut the basin in its final stretch, before the channel excavated in a natural depression (Dal Darra) that carried the waters to the riverbed downstream (Figure 5.11A). Inside the stilling basin the rock consisted of layers folded and with limestone faults and phyllite little resistant to erosion. Other geological faults occurred in the basin, as shown by Lowe (1973).

The service spillway operated for over ten thousand hours during the years 1975 and 1978, with a maximum flow of 8.900 m^3/s. The erosive process occurred rapidly.

During the initial period in 1975, during the first two weeks, a crater was opened in the bed 30m deep, up to the El. 335,00 m; in a large area on the right side of the basin, erosion reached 20 m depth. The igneous intrusion was unchanged. The right side eroded much more than the left, as can be seen in Figures 5.11H.

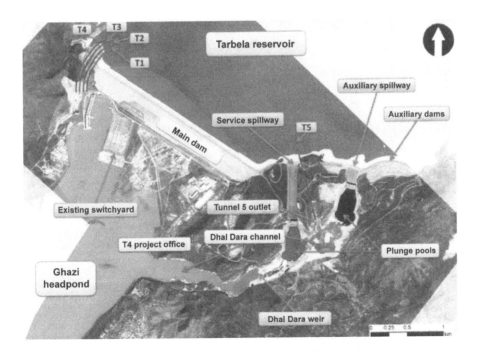

Figure 5.11 Tarbela HPP arrangement.

Figure 5.11A Tarbela HPP. Side view.

Figure 5.11B Tarbela HPP. Side view. Auxiliary spillway working.

Figure 5.11C Auxiliar spillway; (nine gates, 24.070 m³/s).

Figure 5.11D Service spillway; (seven gates, 18.400 m³/s). Observe rocky massif.

Figure 5.11E Service spillway. Repair services.

In 1973/74 a protective wall was built to contain the erosive process, with drainage galleries behind the wall. In addition, to try to stabilize the massif were carried out rock anchors, but without success, the erosion continued. Three years later, in 1977/1978, more rock anchors and a wall of buttresses were built in front of the one previously built. A concrete groyne was also built. (Figure 5.11I). Reduced model trials indicated that the bottoms of the scour pits had to be lowered to increase the water depth to dissipate the jet energy. The plunge pool was drilled. In addition, more rock anchors were applied. In total, 1.660 rock anchors of 21 m length were installed.

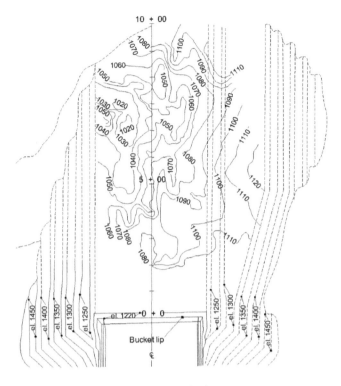

Figure 5.11F Service spillway. Plunge pool – as built.

Figure 5.11G Service spillway. Scour at right side.

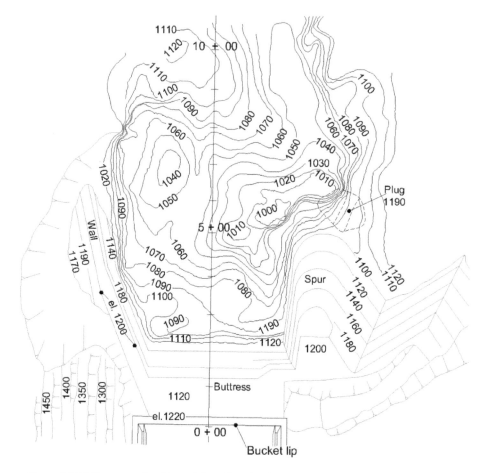

Figure 5.11H Scour repairs. Note: left wall and right spur.

Figure 5.11I Scour containment rock anchors on the downstream slope of the service spillway.

Figure 5.11J Tarbela HPP. Service spillway. Execution of the erosion containment rock anchors on the downstream slope.

The following figures show aspects of the design, erosion and repair work.

5.2.4   *Spillway of Grand Rapids*

The case of Grand Rapids in Canada, 440 MW, was presented by Feldman (1970) (see Figures 5.12 through 5.12E). This plant was completed in 1964 on the Saskatchewan river where it empties into Lake Winnipeg, as shown in the following map.

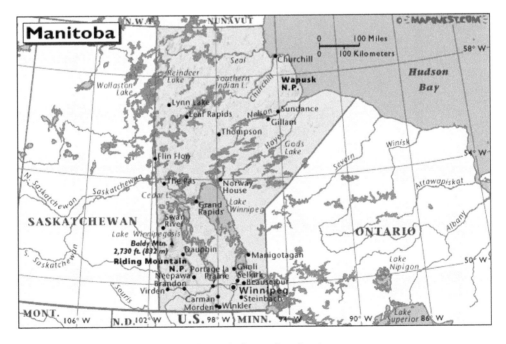

Figure 5.12 Grand Rapids HPP (Canada). *Source:* Localization map.

Figure 5.12A Grand Rapids HPP (Canada). *Source:* Downstream view of the plant.

The arrangement of the works is shown in Figure 5.12BM. The spillway has a discharge capacity of 3.965 m³/s and a drop-off of 18 m (*H*). The takeoff velocity of the jet is of the order of 19 m/s.

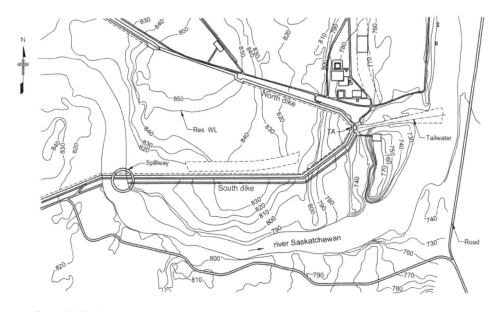

Figure 5.12B Grand Rapids layout.

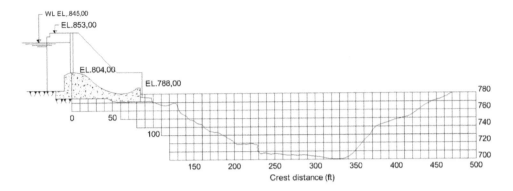

Figure 5.12C Grand Rapids HPP. Spillway profile.

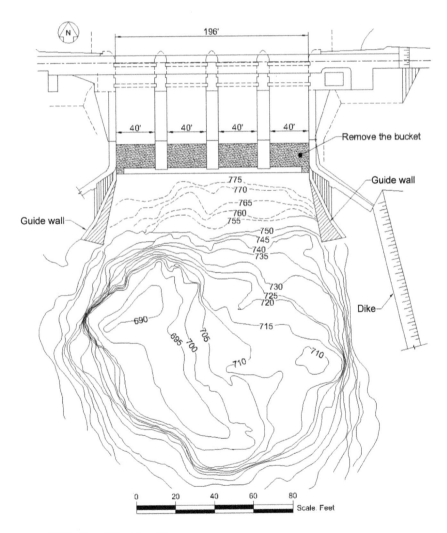

Figure 5.12D Grand Rapids spillway arrangement.

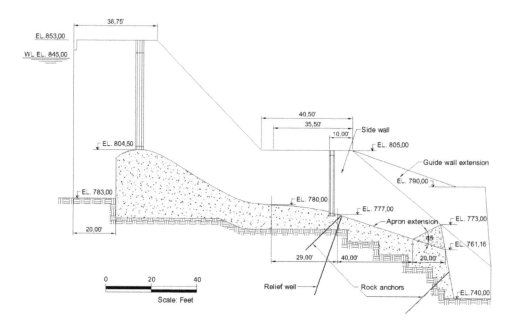

Figure 5.12E Grand Rapids spillway. Solution adopted to contain the erosive process.

The local geology is composed of sandstones and dolomites, in lodging plans, frag-mented and containing layers of clay, silt and sandy material, infrequent. The physical char-acteristics of the rock vary greatly from bed to bed. Two separate joint systems predominate in the jet area and both dives northwestward from each shore. The upper layers are hard sandstone, resistant to scour. The lower layers, however, are susceptible to scour, since they are confined by fragmented thin layers intercepted by vertical joints. The scour pit is shown in Figures 5.12C–5.12E, reaching 24 m deep.

The modification of the spillway to contain the erosive process, studied in a reduced model, is presented in Figure 5.12E.

5.2.5 Spillway of Água Vermelha

There are many cases of erosion downstream of spillways. The return (or recirculation) cur-rents usually affect the lateral areas of the dam slopes and upstream of the impact zone of the jet and may or may not erode them, depending on their intensity and the resistance of the massif, as was the case of the Água Vermelha HPP.

The spillway was designed for a flow of 20.000 m^3/s, with six segment gates of 15,0 × 19,6 m and energy dissipation in a ski jump. Downstream, the top of the rock was at El. 327,70 m and the plunge pool at elevation 314,10 m. On January 20, 1980, with the outflow of 5.750 m^3/s, the return streams caused the beginning of an erosive process of the embankment downstream of the embankment of the dam on the left bank (see details in Oliveira, 1985). In two days, 40.000 m^3 of basalt blocks were removed from the slope. Figure 5.13 shows the wall of protection of the dam on the left side of the spill-way, built between May and December 1980 (see CBDB, 1982a).

Figure 5.13 Água Vermelha HPP.
Source: (CBDB, 1982a).

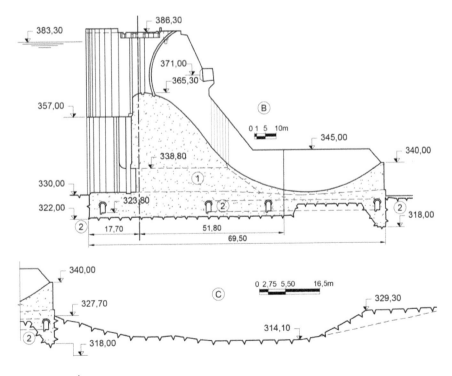

Figure 5.13A Água Vermelha HPP spillway section

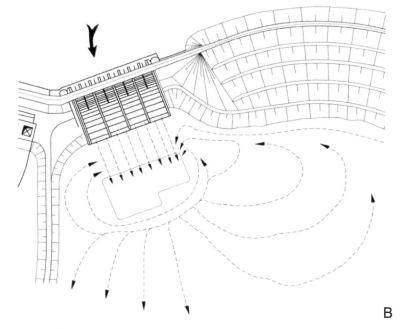

B

Figure 5.13B Água Vermelha spillway. Flow pattern.
Source: (Oliveira, 1985).

Figure 5.13C Eroded area.
Source: (Oliveira, 1985)

In the development of the scour pit, pressure fluctuations cause the dislocation of the blocks in fractured masses. These blocks, recirculating inside the pit, clash and, by abrasion, fragment, then are expelled from the crater, being able to form bars (deposits) immediately at the exit or to be transported downstream.

5.2.6   Spillway of Itapebi

Figures 5.14A and 5.14B show the Itapebi HPP, 450 MW, Brazil. The dam is of the concrete face rockfill type (CFR), with a crest at El. 112,00, a length of 583 m and a maximum height of 121 m. The generation circuit, located in the right bank, is constituted by a water intake of the hollow gravity type, with three pressure tunnels (diameter of 7,40 m and average length of 160 m), supplying the semi-indoor powerhouse, equipped with three vertical shaft Francis turbines.

In July 2001, during the excavation of the spillway, a large rock slide occurred in the left abutment (Figure 5.14C). Due to the unforeseen geological conditions, in order to stabilize the excavation, there was a need to modify the design of the geometry of the chute at its end for a ski jump in the area already excavated for the stilling basin.

For this, a block of RCC was excavated for the ski jump to launch the jet of water further away, thus avoiding the risk of regressive erosions of the rock. In order to permit the

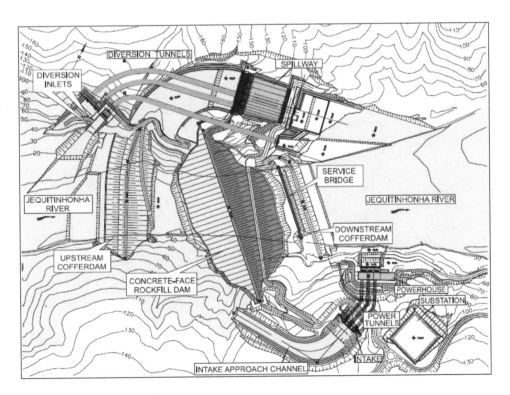

Figure 5.14A Itapebi HPP layout.

Figure 5.14B Itapebi HPP
Source: Downstream view.

Figure 5.14C General view of the left abutment straight after the slide of the rock mass over of the layer of biotite-schist.

adaptation of the ski jump to the already excavated local conditions, the solution presented was a flip bucket inclined 45° with respect to the axis of the chute.

The solution was studied on the hydraulic model. Its operation after various years qualifies it as an adequate solution.

The spillway was designed for a 10.000-year flood of 20.915 m³/s, constituted of six bays with radial-type gates, 17,40 m wide and 21,17 high.

It should be noted that Xavier (2003) presented a paper about this project entitled "Unconventional Solution for the Itapebi HPP Spillway," which describes in detail the singular solution adopted for the project's suitability to the geological conditions identified only during the works.

5.2.7 Spillway of Katse dam

Free fall spillways used in arch dams in narrow valleys or jets from orifices are worth the same observations for those in ski jump. In the case of free fall dissipators, due to the reduced throw distance, additional safety measures may be required, such as the construction of submerged sills (tailwater dam) downstream to raise the depth of the water level and increase the protection of the bottom of the river.

This is the case of the Katse dam in the Malibamatso river, Lesotho (Figure 5.15). The dam is part of the Lesotho Highlands Water Project, with the purpose of irrigation and drinking water to export to South Africa (Furstemburg *et al.*, 1991).

The arch dam is 185 m in height and 710 m in length. The head is 140 m. The free fall spillway has a discharge capacity of 6.000 m³/s.

Figure 5.15 Katse dam (Lesotho).

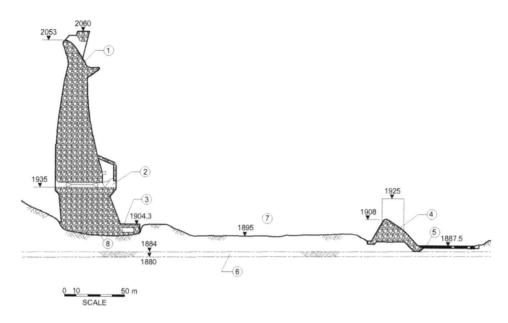

Figure 5.15A Long section through Katse dam.

Source: (Furstenburg *et al.*, 1991).

The main hydraulic structures associated with the Katse dam are: (1) spillway; (2) bottom outlets; (3) mini hydropower station; (4) tailwater dam with a maximum height of 30 m in the central stretch; (5) stilling basin; (6) auto-brecciated layer; (7) plunge pool; and (8) foundation level. Shown in the illustration is dam block number 13 on the

Figure 5.15B Katse 1:30 scale sectional model.

Source: (Furstenburg *et al.*, 1991).

Figure 5.15C Katse 1:70 scale comprehensive model.

Source: (Furstenburg et al., 1991).

right bank, so that the section is through the bottom outlets; the dam foundation extends down to level 1.880 m over the central of the arch.

To select and optimize the most effective spillway layout and associated tailwater dam, it was decided to build two models, one in France and one in South Africa.

The French model was constructed by Sogreah in a laboratory in Grenoble. It was a sectional model to a scale of 1:30, designed to optimize the spillway arrangements for both options and to identify the most effective arrangement. The model included a movable bed to facilitate assessment of the relative scour for different spillway configurations. Figure 5.15B shows the model arrangement.

The South African model was constructed in the laboratory of the Division of Earth, Marine and Atmospheric Science and Technology of the CSIR at Stellenbosch. This comprehensive model was built to a scale of 1:70 and was designed to study scour in the plunge pool and to facilitate sizing of the tailwater dam and its stilling basin. The model included the arch dam, the arch spillway, some 600 m of upstream river and about 1.800 m of the river downstream of the arch as well as the tailwater dam. The model included moveable bed facilities. Figure 5.15C shows the model layout.

After optimization of the bucket spillway layout to achieve maximum flow dispersion and aeration, it was necessary to optimize the tailwater height such that the scour would not proceed below El. 1.888 m. This was the aim of the comprehensive model built in South Africa.

The tail water height was optimized to safeguard the dam against floods up to the routed probable maximum flood of 6.000 m^3/s. The final arrangement and the typical scour pattern are shown on 5.15D.

To reduce scour from the bottom outlet jets to acceptable levels, it was necessary to rotate the outlets horizontally, one a few degrees more than the other, so that they formed separate, shallower scour holes rather than one large hole.

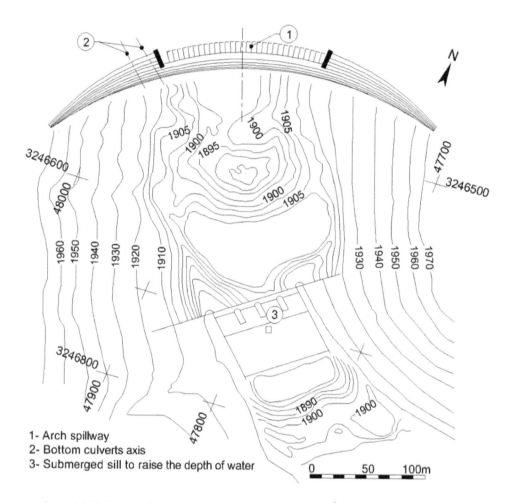

1- Arch spillway
2- Bottom culverts axis
3- Submerged sill to raise the depth of water

Figure 5.15D Katse spillway. Scour pattern after Q = 6.250 m³/s.

Source: Furstenburg *et al.* (1991).

Foundation conditions influenced the design of both the spillway and the plunge pool. Weak heterogeneous auto-brecciated layers at relatively shallow depth below the riverbed gave cause for concern. Exposure of these layers to erosion and to rapidly pulsating water pressure could adversely affect the stability of the dam and its abutment. It was therefore essential to ensure that the brecciated layers would remain divorced from turbulence and pressure fluctuations resulting from the overspilling flow.

5.2.8 *Spillway of Kariba*

Figure 5.16 shows the evolution of the erosive process downstream from the spillway of the Kariba power plant (1.626 MW) on the Zambezi river, Zambia/Zimbabwe, Africa (Figure 5.16). The dam, with a height of 128 m, was seated at El. 362,00 m. The spillway has six orifices, 9 m × 9 m each, with flow capacity of 1.400 m³/s.

The local geology is composed of sandstones, conglomerates, sand/gravel (from the Triassic), clayey strata (from the Permian), quartzite (from Precambrian), gneisses, amphibolites and schists (from the Archaean). The riverbed and the abutments are formed by very stable gneiss. On the right abutment, 90 m above the riverbed, the gneiss is covered by a quartzite with joints and a strong foliation (Anderson, 1960). On both abutments, the materials were heavily weathered.

The plant started operation in 1960. The erosion process evolved rapidly, as shown in the following figure. The erosion pit already had, in 1981, 80 m of depth. In 2001, it remained the same elevation according to Noret (2012).

Noret (2012) presented a summary of the various estimates made for the depth of the Kariba pit over the years. All were underestimated: Veronese (1937) 318 m; Damle (1966) 359 m; Martins (1975) 359 m; Chee and Kung (1974) 336 m; Sofrelec (1980) 323 m and Mason (1986) 313 m. The modeling of the phenomenon was improved by the calibration of a formula developed specifically for Kariba by Alain Carrère and by Mason (Coyne et Bellier, 1997–1998).

$$D_{max} = \frac{5,27 \; q^{0,61} H^{0,05} h^{0,15}}{g^{0,30} d_m^{0,10}} \tag{5.3}$$

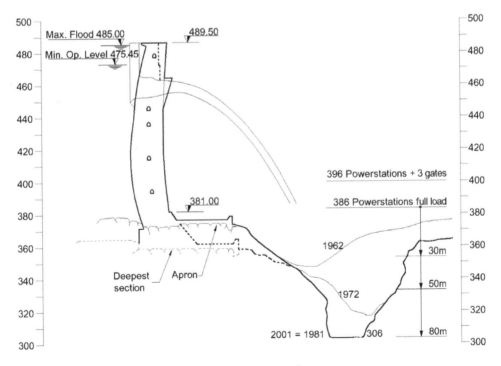

Figure 5.16 Kariba dam. Orifice spillway. Q = 8.400 m³/s. Scour evolution.

Source: (Noret, 2012; CBDB, 2002; Whittaker and Schleiss, 1984; Mason, 1986; Hartung and Häusler, 1973).

where:

D_{max} = maximum depth of scour calculated using this formula (m);

H = total head (m);

h = downstream depth of water (m);

g = acceleration of gravity (m^2/s);

dm = mean particle diameter (m).

According to Noret (2012), for design discharge of 8.400 m^3/s, with the six gates open, scour reached the elevation 293,00 m. Noret reports that a modern bathymetric survey carried out with the support of divers in March 2011 revealed that there are no significant changes in the geometry of the pit compared to 1981 (Figure 5.17). It also reports that the spillages during this period were rare and that the spillways with more than three gates were avoided.

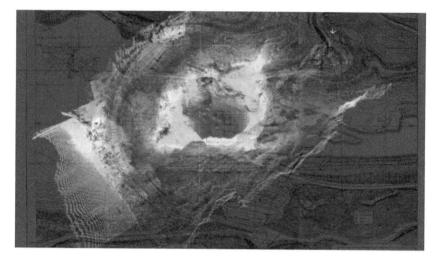

Figure 5.17 Bathymetric survey conducted in March/2011.
Source: (Noret, 2012).

A study of remodeling the plunge pool was followed. The Zambezi River Authority contracted the Hydraulic Construction Laboratory of the Federal Polytechnic School of Lausanne (LCH/EPFL) and Aquavision Engineering of Switzerland, to evaluate the future evolution of the scour and to study a new form of the pit that could guarantee the safety of the dam.

The studies were carried out using the mathematical model CSM – Comprehensive Scour Model (Bollaert, 2010, 2012) and a fixed bottom physical model in the 1:65 scale. The pressures measured in the physical model supplied the mathematical model. The hydrodynamics pressures were compared with the strength of the rock mass (gneiss on the left abutment and a quartzite weathered on the right abutment). The analysis of the results allowed to conclude that the current form of scour was unstable for the condition of adjacent gate operations, including a regressive erosion tendency toward the dam foot (see Figure 5.18).

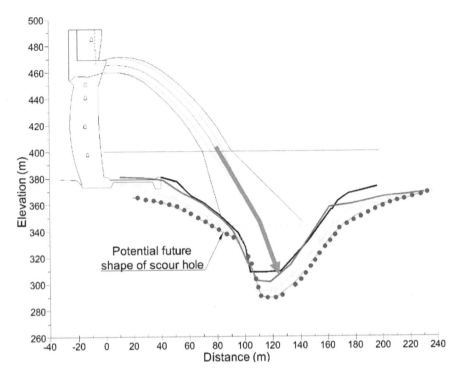

Figure 5.18 Future potential erosion downstream of the Kariba dam.

Source: (Bollaert, 2012).

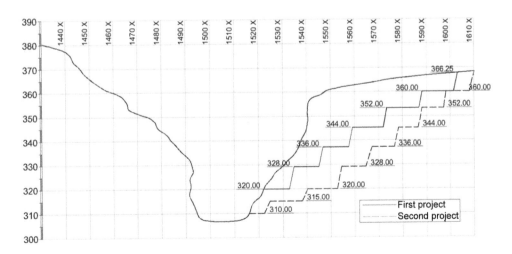

Figure 5.19 Geometry of the Kariba plunge pool re-excavation project.

Source: (Noret, et al., 2012).

As a result, the Zambezi River Authority has decided to contract the remodeling services of the erosion form to continue the operation of the plant in a safe manner.

Services in the process of contracting (October 2012) involved additional excavation to alter the shape of the downstream slope of the erosion pit (plunge pool), making it smoother, as well as spillway modernization services in separate contracts, according to the studies of Noret (2012) and Bollaert (2012).

The flooding period runs from February to April. The excavation work, with an estimated volume of $3{,}0 \times 10^5$ m^3, should then be carried out in nine months from May to January, with both powerhouses operating at full load.

The area will be isolated by cellular cofferdams, 20 m high, filled with random-granular materials.

Excavations should be careful given the proximity of the dam structure, the outlets of the powerhouses and the unstable south bank.

Figure 5.20 Kariba HPP (1.626 MW).

Figure 5.21 Kariba HPP. Spillway running.

5.2.9 Spillway of Cahora Bassa

It is worth mentioning the spillway project at the Cahora Bassa HPP, on the Zambezi river, downstream from Kariba, with a spillway in crossed jets.

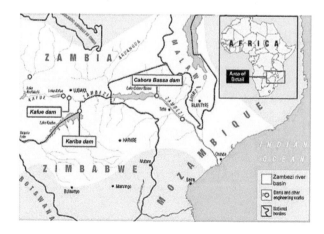

Figure 5.22A Cahora Bassa HPP; 2.000 MW. Zambezi river. Mozambique.

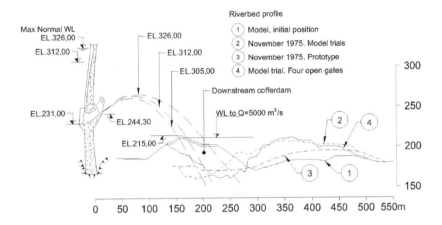

Figure 5.22B Cahora Bassa HPP. Riverbed profiles. Spillway in crossed jets. $Q = 13.100$ m³/s. Dam height = 170 m.

Source: (Lemos, 1983).

Figure 5.23 UHE Cahora Bassa. Mozambique. Spillway in crossed jets.

Source: (Lemos and Quintela, 1982).

5.2.10   Spillway of Itaipu

Reference should also be made to the Itaipu HPP spillway. The hydraulic behavior after eight years of operation was presented by Szpilman, et al. (1991).

The excavation project predicted the slope shown in the Figures 5.24 to 5.26. From the base, in the El. 95,00 m, through the berm, El. 115,00 m, and up to the elevation 120,00 m, appears the dense basalt flood B, slightly fractured (F1 and F2), with excellent geotechnical conditions.

From there up to El. 133,00, breach B appears of good quality, but with isolated pockets of less resistance, type III and IV.

The layer C caps the basaltic mass. No discontinuity of greater significance was detected in the area. Downstream of the chutes, the outcropping rock is composed only of dense basalt flood B.

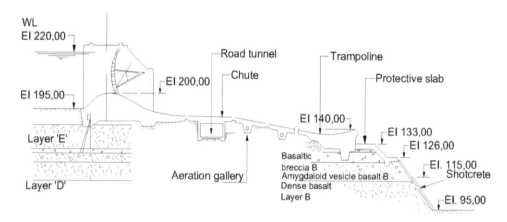

Figure 5.24 Spillway of the Itaipu HPP.
Source: (Szpilman, et al., 1991).

Figure 5.25 Itaipu HPP. Spillway in 1982, before the operation.
Source: (CBDB, Large Brazilian Spillways, 2002).

Figure 5.25A Rocky slope of basalt on the right bank, downstream of the spillway (1987).

Figure 5.26 Itaipu HPP. First flood in the spillway (1983).

Source: (CBDB, Large Brazilian Spillways, 2002).

The hydrodynamics pressures resulting from the continuous operation of the spillway during the first four years, with exceptional flows that reached almost 40.000 m^3/s, caused erosions, as shown in Figures 5.27 through 5.33.

Figure 5.27 Itaipu HPP. Scour observed in the right lateral wall of basalt downstream of the spillway in contact between vesicular basalt and dense columnar basalt – 1987, after four years of operation.

Source: (Personal file).

All floods that occurred during the first 20 years of operation, as well as the respective massive reinforcement repairs (rock anchors, building blocks, walls), are recorded in the

Figure 5.28 Itaipu HPP. Scour observed in the right lateral wall of basalt downstream of the spillway in 1987.

Source: (Personal file).

Figure 5.29 Itaipu HPP. Start of repair work − 1987.

Source: (Personal file).

Figure 5.30 Itaipu HPP. Scour downstream of the spillway already repaired (1987).

Figure 5.31 Itaipu HPP. Scour downstream of the spillway already repaired (1987).

Figure 5.32 Itaipu HPP. Region downstream of the spillway already repaired (1987).

Figure 5.33 Itaipu HPP. Region downstream of the spillway already repaired (1987).

Figure 5.34 Itaipu HPP. Downstream region of the spillway October 2014.

Figure 5.35 Itaipu HPP. Downstream region of the spillway October 2014.

CBDB publication, "Large Brazilian Spillways" (2002). The photos in Figures 5.34 and 5.35, provided by Eng. Dimilson Pinto Coelho of Itaipu, show the current situation of the region in October 2014.

5.2.11   *Spillway of Itá*

Itá HPP (see profile in Figure 5.36) has two spillways: one main with six gates, the other with four gates, all 18 m wide × 20 m high, as shown in Figure 1.9. The discharge capacity is 50.000 m^3/s.

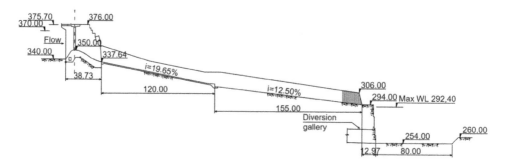

Figure 5.36 Itá HPP. Spillway channel 2, profile.

Source: (CBDB, 2009). Plant – see Figure 1.9.

The velocity of the flow at the launching edge of the spillway jet 2, elevation 294.00 m, is on the order of 39 m/s.

Figures 5.37A–F show the erosion process that occurred in spillway 2 of the Itá HPP, courtesy of Engineer Sergio Correa Pimenta.

Figure 5.37 A through D Itá HPP. Scour downstream of spillway channel 2. Situation on June 28, 2014.

Figure 5.37-E Itá HPP. Scour downstream of spillway channel 2. Situation on June 28, 2014.

Figure 5.37-F Itá HPP. Scour downstream of spillway channel 2. Side view. Situation on June 28, 2014.

5.2.12 *Spillway of Fengtan*

This project was presented by Junshou at the International Symposium on General Arrangements of Dams in Narrow Vales, Rio de Janeiro (CBDB/ICOLD, 1982). The Fengtan HPP (800 MW) is located on the Yoshui river, Zhangjiabe, Hunan Province, China. Is a hollow gravity dam 112,50 m high and 488 m long.

Figure 5.38 Fengtan HPP. China. Cross-section of the dam, spillway and powerhouse. See also Figure 2.1C.

Source: Junshou MA (CBDB/ICOLD, 1982).

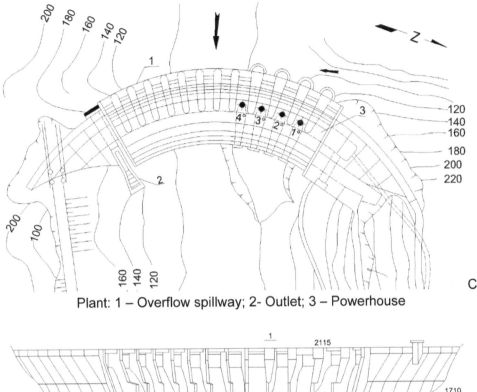

Plant: 1 – Overflow spillway; 2- Outlet; 3 – Powerhouse

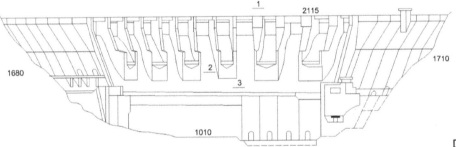

Downstream view: 1 – Overflow spillway; 2 – High deflecting bucket; 3 – Low deflecting bucket.

Figure 5.38 (Continued)

The local geology is composed of quartz sandstones, with clayey layers intercalated, dipping 45° downstream, with faults, joints and fissures and deeply altered rock, with little resistance to scour. In addition, the depth of water downstream to protect the massif was small. This aspect was studied exhaustively in a reduced model.

The spillway, which incorporates the powerhouse, has a discharge capacity of 23.300 m³/s. It was designed with 13 spans, with gates 14 m wide × 12 m high, with tall buckets in six spans and low buckets in seven spans – 28 m below. In addition, a bottom outlet was designed on the right abutment, with a 6 m × 12 m sluice gate for emptying the reservoir. The specific discharge is of the order of 183 m³/s/m. The structure operated frequently from 1970 to 1980. The maximum discharge was 12.500 m³/s in the 1980 flood. The survey showed that a scour of 19 m depth was developed, according to what was foreseen in the reduced model. The author did not obtain updated data on the evolution of erosion.

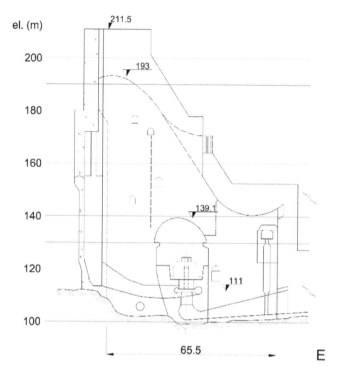

Figure 5.38   (Continued)

5.3   Hydraulic jump dissipators

5.3.1   *Summary of theory*

The hydraulic jump is defined as the abrupt and turbulent passage of the flow from a low energy stage, below the critical depth, to a high energy stage, above the critical depth, during which the velocity changes from supercritical to subcritical – Figure 5.39 (HDSBED, 1983).

   That is, the hydraulic jump is the process of dissipation of the kinetic energy of the supercritical flow in potential energy in the subcritical flow in the stilling basin. The jump is accompanied by a violent impact and consists of an abrupt increase of the water level in the region of the impact between the fast flow and the slow flow.

Definition of symbols:

- Q = total discharge (m³/s); W = width of the basin (m); $Q/W = q$ = unit discharge (m/s/m);
- $d_c = \sqrt[3]{q^2/g}$, is the critical depth;
- V_1, d_1 and E_1 = velocity (m/s), depth (m) and energy in Section 1;
- V_2, d_2 and E_2 = velocity (m/s), depth (m) and energy in Section 2;
- $V_1 = \sqrt{2g(Z - H/2)}$, where $Z = WL_{res}$ – elevation of the basin; H = head on spillway (m);
- $F1 = V_1(gd_1)^{0.5}$; L = length of the jump (m).

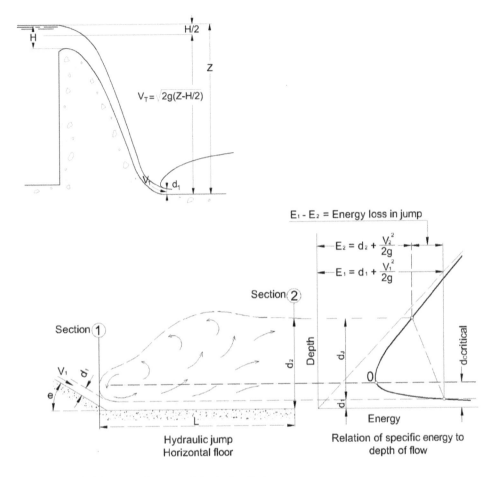

Figure 5.39 Schematic diagram of the hydraulic jump.

Source: (HDSBED-USBR, 1983).

A hydraulic jump will form when a supercritical flow with small water height strikes a body of water with a considerable height and subcritical velocity. It should be noted that the jump is formed only if the upstream height is less than the critical height.

For the jump to form, the pressure plus the momentum after the jump should be equal to the pressure plus the momentum before the jump. The pressure plus momentum requirements remain valid regardless of the shape or slope of the channel.

The stilling basins are presented by Elevatorski (1959), Rudavsky (1976) and Peterka (1983). The following is a summary only, considering the schematic diagram of the hydraulic jump of Figure 5.40. In the analysis, the following hypotheses are made:

* the channel is rectangular, with parallel sides, and the bottom slab is horizontal;
* all friction losses are neglected;
* it is assumed that the jump happens instantly;
* there is a streamline flow upstream and downstream of the jump.

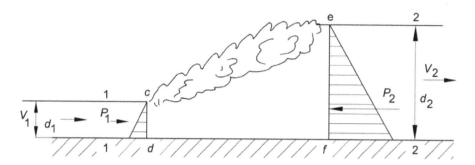

Figure 5.40 Schematic diagram of a hydraulic jump.

Source: (Elevatorski, 1959).

P_1 is the hydrostatic pressure in Section cd, and P_2 the hydrostatic pressure in Section ef. It is assumed, for simplification, that the channel width is unitary. The variation in hydrostatic pressure ΔP from Section 1 to Section 2 for a channel of unit width is:

$$\Delta P = P_2 - P_1 = \frac{1}{2}\gamma(d_2^2 - d_1^2) \tag{5.4}$$

The momentum variation is:

$$\Delta M = \frac{\gamma q(V_1 - V_2)}{g} \tag{5.5}$$

According to Newton's Second Law, the variation of the hydrostatic pressure is equal to the momentum variation. Equating $\Delta P = \Delta M$, we have:

$$\frac{\gamma q V_1}{g} + P_1 = \frac{\gamma q V_2}{g} + P_2 \tag{5.6}$$

For rectangular channels the equation becomes:

$$d_2^2 - d_1^2 = \frac{2q}{g}(V_1 - V_2) \tag{5.7}$$

As reported by HDSBED (1959), several researchers worked on the subject, until one reaches the classic dimensionless equation presented in Equation 5.8:

$$\frac{d_2}{d_1} = \frac{1}{2}(\sqrt{8F^2 + 1} - 1) \tag{5.8}$$

It has been proven that the efficiency of a stilling basin is a function of the kinetic flow factor, Froude (F_1) number at basin inlet and basin type. This factor, dimensionless, is given by Equation 5.9, where V_1 and d_1 are the velocity and the height at the entrance of the basin, respectively, and g is the acceleration of gravity (9,81 m/s^2).

$$F_1 = \frac{V1}{\sqrt{gd_1}} \tag{5.9}$$

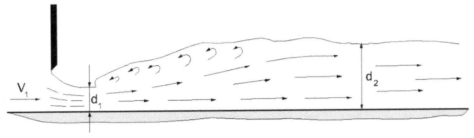

F_1 between 1,7 and 2,5
Form A - prejump stage

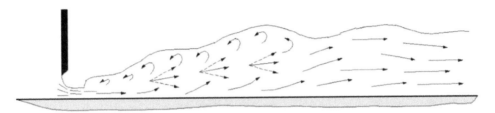

F_1 between 2,5 and 4,5
Form B - transition stage

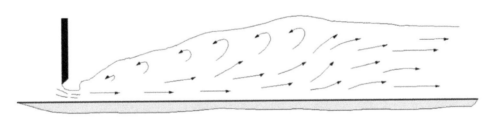

F_1 between 4,5 and 9,0
Form C - range of well-balanced jumps

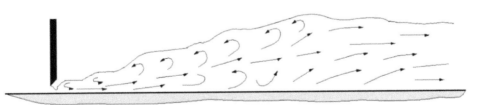

F_1 greater than 9,0
Form D - effective jump but rough surface downstream

Figure 5.41 Types of hydraulic jumps (function of the Froude number).
Source: (HDSBED, 1983).

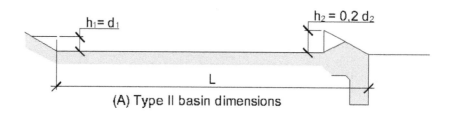

(A) Type II basin dimensions

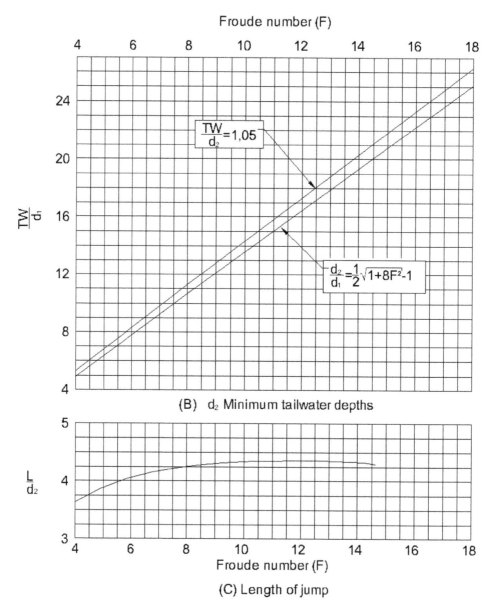

Froude number (F)

$\dfrac{TW}{d_2} = 1,05$

$\dfrac{d_2}{d_1} = \dfrac{1}{2}\sqrt{1+8F^2}-1$

(B) d_2 Minimum tailwater depths

(C) Length of jump

Figure 5.42 Basin Type II.
Source: (HDSBED, 1983; Lysne-NIT, 2003).

As a function of F_1, there are four different forms of hydraulic rebound shown in Figure 5.41 (HDSBED). All these forms are found in practice.

Figure 5.42 shows a Type II basin and the graphs used for hydraulic design, its dimensions as a function of the Froude number and the conjugated heights of the jump (d_1 and d_2). It is noteworthy that, as F_1 increases, the efficiency of the basin also increases. Otherwise, efficiency is greater as the height of the shoulder ($d_2 - d_1$) increases, with d_2 being the height at the exit of the basin.

The length of the basin (L) is a function of the height of the shoulder. Elevatorski (1959) recommends when the bedrock is fractured:

$$L \cong 7(d_2 - d_1) \tag{5.10}$$

and 60% of this value when the riverbed is made up of sound rock.

The parameters d_1, d_2 and L are shown in Figures 5.41 and 5.42. From the practice of projects, it is known that basins with more than 30 to 40% efficiency are not common. It follows that 70 to 60% of the energy entering the basin leaves and must be dissipated in the mass of water and in the rocky massif downstream.

5.3.2   Stepped spillway of Dona Francisca

One of the ways to increase the dissipation of kinetic energy before the stilling basin is to build steps along the spillway profile, as shown in Figure 5.43, which leads to a reduction of the residual energy at the entrance of the basin (see also sections 2.11 and 3.6.4). The energy dissipated by the steps can be significantly reduced, which will mean savings in the basin project.

An example of this type is the case of the Dona Francisca HPP spillway, designed for a discharge capacity of 10.600 m³/s (Figures 3.32A and 3.32B). The length of this stilling basin was reduced by 14%, from 50 m to 43 m, as reported by Sobrinho (2000).

Interest in this type of solution increased with the advent of CCR dams in the 1960s. The process of running the compressed concrete layers allows the execution of the finish in steps with great ease (Figure 5.43).

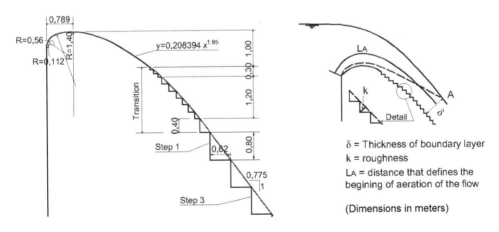

Figure 5.43 Typical configuration of a stepped spillway.
Source: (CBDB, 2002).

The hydraulic performance of this type of spillway was studied in models by Tozzi (1992). There are several papers available on the internet about the subject matter; Chanson (1993, 1994, 1995, 2001), Sanagiotto (2003) and Simões (2008) are recommended. An "Investigation on Stepped Spillways" is also recommended, a summary presented in the Main Brazilian Spillways book – CBDB/LBS (2002).

5.3.3   Spillway of Porto Primavera

Another example of a hydraulic jump stilling basin is the spillway of the Porto Primavera HPP (Figure 5.44) on the Paraná river, with a discharge capacity of 52.800 m³/s. The project was started in 1980, and the plant was inaugurated in 2003.

Figure 5.44 Porto Primavera HPP (1.568 MW), Paraná river, São Paulo/Mato Grosso.
Source: CESP. (CBDB/MBD, 2009).

It has a 10,4 km long earth dam and a powerhouse 558 m long, including the assembly area, 14 Kaplan turbines of 112 MW and design head of 18,3 m.

The spillway is 315 m long, with 16 gates 15 × 18 m and a discharge capacity of $Q = 52.800$ m³/s; energy dissipation is in a 35,2 m long hydraulic bucket (roller bucket). Observe the flow pattern.

It should be noted that the classical back currents often bring blocks loose by the action of the flow into the basin, which causes erosion in the concrete by abrasion. This fact is almost always aggravated during asymmetric operations of the spillway gates. There are several other examples presented later in this book.

5.3.4   Spillway of Ilha Solteira

Ilha Solteira, Paraná river, is an earth dam with a length of 7,2 km. The final design was carried out from 1969 to 1974, when the generation started. The powerhouse has a total length of 633 m, including the assembly area, 20 Francis turbines of 160 MW and design head of 46 m.

The spillway has a total length of 351,50 m, 19 gates of 15 × 15 m, and a total discharge capacity of 40.000 m³/s. The energy dissipation is in a hydraulic jump stilling basin. Baffle blocks were inserted in the descending slope of the Creager profile. The stilling basin is 71,5m long. Downstream of the end sill, a complementary excavation was made to allow the expansion of the flow and reduce the possibilities of scour.

Figure 5.45 Ilha Solteira HPP (3.200 MW).

Source: CESP. (CBDB/highlights, 2006).

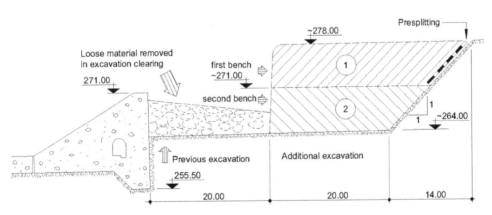

Figure 5.45A Complementary excavation to eliminate abrasive actions.

Source: (Neidert, 1980a, 1980b).

5.3.5 *Spillway of Estreito*

An erosion pattern downstream of the stilling basins is the case of Estreito HPP, on the Tocantins river (Figure 5.46), which was studied in a reduced model in the FCTH-USP. Another view of this erosive pattern is shown in figure 10.3. The zones of erosion and deposit and formation of bars are characteristics.

Figure 5.46A–B Estreito HPP. Tests in hydraulic model with movable bottom. Erosion pattern downstream of stilling basin. Q = 28.700 m³/s.

Source: (Fudimori, 2013).

Figure 5.46C Estreito HPP, Tocantins river. Spillway; some openings work. (1.087 MW; head = 22 m. 8 Kaplan turbines) (April 2011).

5.3.6   *Spillway of Coaracy Nunes*

It is also worth mentioning the erosions downstream of spillways greater than those estimated in the project due to:

*   their under-dimensioning, or of stresses arising from high negative pressures resulting from asymmetric sluice gate operations, not provided for by not having a model for studies;
*   the inadequate assessment of rock mass resistance due to the absence or insufficient quantity of geological and geotechnical investigations in the region downstream of the spillway.

Examples of these occurrences are the scours downstream of the spillway of the Coaracy Nunes HPP – 12.000 m³/s (Figures 5.47A and B) in the Araguari river (State of Amapá), and the bottom spillway of the Moxotó HPP – 28.000 m³/s (Figure 5.48) in the São Francisco river (State of Pernambuco), reported by Neidert (1980a, 1980b) and Bobko (1980). The case of the Sobradinho HPP, also in the São Francisco river, will be presented in Figures 5.49 to 5.52.

Figure 5.47A Coaracy Nunes HPP.
Source: Eletronorte (1975).

The arrangement contemplates a dam and a spillway in the left bank and another dam and a powerhouse (78 MW) in the right bank. It should be noted that after 39 years, Eletronorte reviewed the hydrological studies and the new 10.000 year flow was reduced to 7.213 m³/s.

The spillway had been designed without the aid of hydraulic model studies. After the start of the operation, the erosive process occurred (Figure 5.47C). Eletronorte then hired CEHPAR to study the model case. As a solution, a new terminal sill was introduced (see Figure 5.47D), and changes were made in the operating plan of the gates.

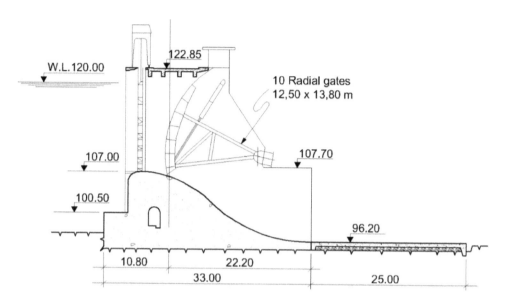

Figure 5.47B Spillway of the Coaracy Nunes HPP (Eletronorte).

Figure 5.47C Erosion in the stilling basin, for $Q = 4.000$ m³/s.

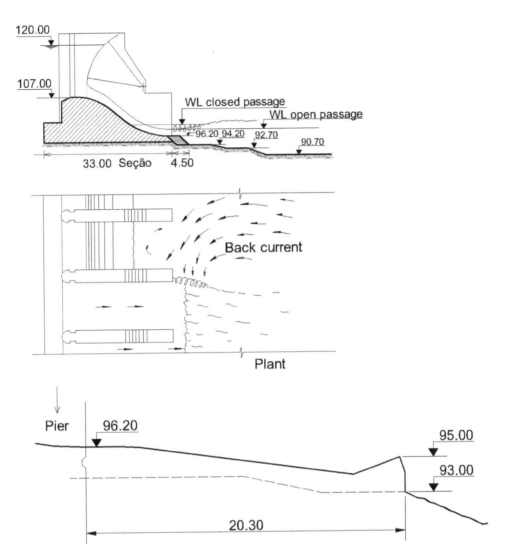

Figure 5.47D Spillway of the Coaracy Nunes HPP. Built alternative. Asymmetric operation scheme. Scour.

Source: Alternative Solution Studies. CEHPAR (Bobko, 1980).

5.3.7 Spillway of Moxotó

Moxotó UHE, completed in 1974 (Figure 5.48A), is part of the Paulo Afonso Hydroelectric Complex. Moxotó is composed of two rockfill dams separated by an island. One of the dams incorporates the intake and the powerhouse with four 100 MW turbines (Figure 5.48 B), and the other one incorporates a culvert spillway (bottom outlet), 20 gates, each 10 m × 8 m, to a discharge capacity of 28.000 m^3/s (Figure 5.48C).

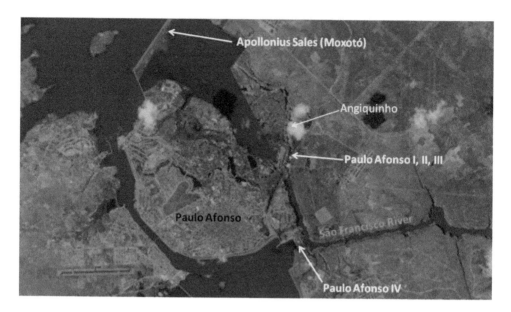

Figure 5.48A Moxotó HPP and the Paulo Afonso Complex.

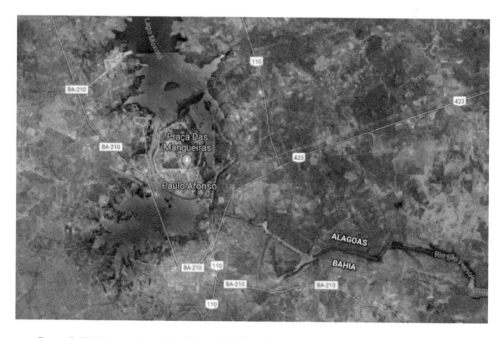

Figure 5.48B Reservoirs of the Moxotó HPP and of the Paulo Afonso Complex.
Source: Sao Francisco river – Google.

Figure 5.48C Moxotó HPP.

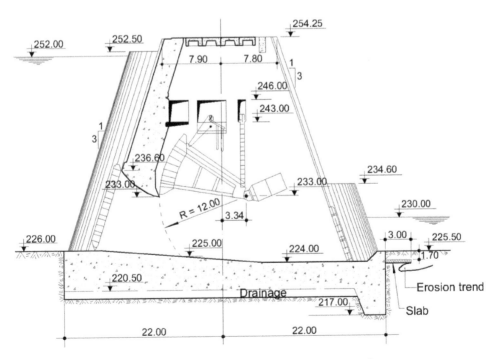

Figure 5.48D Moxotó culvert spillway (bottom outlet); (Q = 28.000 m³/s; 20 gates 10 m × 8 m).

5.3.8 Spillways of Sobradinho

Sobradinho HPP, in the São Francisco river, Bahia, Brazil, owned by CHESF, is composed of a right bank dam, an intake/powerhouse, two spillways and a left bank dam, as shown in Figure 5.49. The spillways are a surface spillway with a discharge capacity of 6.855 m^3/s and a culvert spillway (bottom outlet) of 16.000 m^3/s.

Figure 5.49A Sobradinho HPP – 1.050 MW.

Source: São Francisco river.

The local rocky massif is predominantly composed of gneiss, schist and quartzite, with the quartzite detaching and forming the abutments. The schistosity is parallel to the alignment of the dam axis, dipping 70–80° upstream. The rock at the site is sound of excellent quality, but fractured and therefore without resistance to erosion.

In 1980, downstream of the bottom outlet, an intense erosive process occurred that was predicted in the reduced model studies. Scours higher than predicted in the design can be caused by two factors:

* incomplete knowledge of the key curve, especially for high flows; and,
* the partial operation of the spillway, so that, for a determined concentration of energy, there are no downstream levels capable of counterbalancing it and maintaining the hydraulic jump within the basin.

Figure 5.49B Sobradinho HPP. Culvert spillway (bottom outlet). 1 – Downstream view. Note a built-in cofferdam to enable repairs to be performed. 2 – Culvert spillway in operation.

Source: Photo taken at the plant visit August 1982.

Figure 5.49C Sobradinho HPP. View of the erosion downstream of the culvert spillway (bottom outlet). Note scar of the large diameter block removed.

Source: Photos taken at the plant visit August 1982.

Figure 5.50 UHE Sobradinho. Loose granite-gneiss block downstream of the end sill. Note size of the block, comparing it to the people. In the background is the dividing wall with the surface spillway.

Figure 5.50A Sobradinho HPP. Dividing wall between the bottom outlet and the surface spillway.

Figure 5.51 Sobradinho HPP. Scour downstream culvert spillway close to dividing wall. Behind the wall is the surface spillway. Observe scar of blocks torn off, with diameters larger than people.

Source: (August 1982).

Figure 5.52 Sobradinho HPP. Downstream photo of the culvert spillway (bottom outlet). Note the bottom radial gates. The photographer stands next to the foundation of the dividing wall.

Source: (August 1982).

In operations with a reduced number of gates, it is very common to request the corresponding section of the structure and the rocky massif to conditions much more adverse and more frequent than those corresponding to the flood of design itself.

5.3.9 *Spillway of Grand Coulee*

In the case of roller buckets, which operate similarly to the hydraulic jump, some problems of abrasion of the bucket by materials plucked from the riverbed have been recorded, as at Grand Coulee dam on the Columbia river, reported by Rudavsky (1976). The geology of the place is composed of basalt and granites. The plant (6.480 MW), built between 1933 and 1942, has the following purposes: irrigation, flood control and water supply.

The WL of the reservoir is at a height of 393,20 m. The fall is 105 m. The dam has a maximum height of 168 m; the downstream face of the structure is 0,8 H:1,0 V. The spillway is 503 m wide, 11 gates 41,10 m wide and 8,53 m high. The discharge capacity is 28.315 m³/s (Kollgaard and Chadwick, 1988).

In the body of the spillway, there are 40 outlet works of 2,60 m in diameter, with a capacity of 15.430 m³/s, which discharge into the dissipation basin. The bucket is at the height of 265,14 m. The head is 128,06 m and the flow velocity 50 m/s. The 20 diversion conduits were plugged.

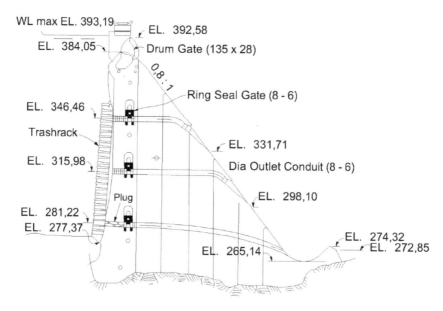

Figure 5.53 Grand Coulee spillway cross-section.
Source: (Kollgaard and Chadwick, 1988).

Figure 5.54 Grand Coulee HPP. Columbia river. Coulee City, Washington – USA.
Source: (Kollgaard and Chadwick, 1988).

5.4   Cases

The publication "Great Brazilian Spillways – An Overview of Brazilian Practice and Experience in Projects and Construction of Spillways for Large Dams," is an important work developed by the CBDB (2002) about the subject matter, which presents the cases of the spillways of:

- Machadinho HPP, on the river Pelotas affluent of the river Uruguay;
- Marimbondo HPP, Camargos HPP, Estreito HPP, Jaguara HPP, Peixoto HPP and Porto Colombia HPP, all on the Grande river in the river Paraná basin, on the border of the states of Minas Gerais and São Paulo;
- Cachoeira Dourada HPP and São Simão HPP, on the river Paranaíba affluent of the river Paraná;
- Itaipu HPP, river Paraná;
- Foz do Areia HPP, river Iguaçu affluent of the river Paraná;
- Itá HPP, Salto Caxias HPP and Salto Santiago HPP, on the river Uruguay affluent of the river Paraná;
- Três Marias HPP and Xingó HPP, on the river São Francisco;
- Orós dam, river Jaguaribe;
- Tucuruí HPP, river Tocantins.

In addition to these cases, it is noteworthy that this publication presents two special articles: one on "Research on Spillways in Steps" and another on "Erosion Downstream of Spillways."

Some of these spillways were briefly presented in this book: Foz do Areia, Itá, Itaipu, Jaguara and Tucuruí. For other cases, it is recommended to consult the publication. Particularly noteworthy are the cases of:

- Machadinho spillway, which has a stretch of 180 m of unlined chute, aiming to save concrete; the flow goes directly on the rocky massif (basaltic floods), like the spillways of Nova Ponte and Serra da Mesa, both presented in this book; and
- Maribondo spillway, which had its basin scoured by blocks of rock brought downstream by the back currents.

Given the importance of the subject, it was deemed appropriate to restate some illustrations/data of these cases in this book. In addition to these are the cases of Nova Ponte and Paulo Afonso IV, whose spillways have chutes in concrete/rock (lined/unlined).

In order to complete the cases, a summary of the Crestuma dam, in the Douro river, Portugal, is presented. This is a particular case of dam/spillway founded in alluvium (40 meters of sand, gravel and some silt).

5.4.1 Spillway of Machadinho

Design discharge TR = 10.000 years	37.350 m³/s
Probable maximum flood (PMF)	39.750 m³/s
Maximum normal WL	480,00 m
Maximum exceptional WL	484,38 m
Absolute maximum WL	485,36 m
Length of the structure	322,00 m
Number of spans	8
Opening width	18,00 m

Gate height	20,00 m
Chute width	175,50 m
Length of the lined chute	97,65 m
Length of the unlined chute	180,00 m
Chute area	31.590 m^2

This solution with an unlined chute with an area of 31.590 m^2 allows a considerable saving of structural concrete (Figures 5.55 to 5.58).

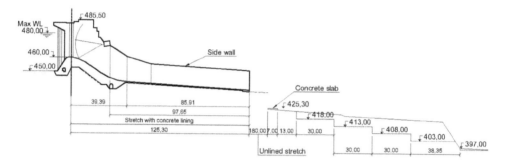

Figure 5.55 Machadinho HPP. Plant and longitudinal section of the spillway.
Source: (CBDB, 2002).

Figure 5.56 Machadinho HPP.
Source: (CBDB, 2002).

Figure 5.57 Machadinho spillway. General view after construction.
Source: (CBDB, 2002).

Figure 5.58 Machadinho HPP. Spillway. First spilling (January 10). $Q = 10.000$ m3/s.
Source: (CBDB, 2002).

Erosions in the rock mass occurred according to what was predicted in the studies in model without risk to the structure. Figure 5.59 shows the bars formed in the riverbed with the eroded material, which caused a loss of 2 m in the head (2% fall in generation).

Figure 5.59 Machadinho HPP. Spillway.Area of the most significant erosion.
Source: (CBDB, 2002).

Figure 5.60 Machadinho HPP. Spillway. Bars formed in the bed of the river with eroded material.
Source: (CBDB, 2002).

5.4.2   *Spillway of Marimbondo*

The 1.400 MW Marimbondo HPP (Figure 5.61) is located on the Grande river (SP/MG) between Água Vermelha HPP and Porto Colombia HPP. Its construction began in 1971, and the first unit was commissioned in 1975 (CBDB, 2002).

Figure 5.61 Marimbondo HPP.
Source: (CBDB, 2002).

The main characteristics of the spillway are:

Capacity	21.400 m³/s
Spillway design discharge	16.000 m³/s (75% of the maximum capacity)
Unit discharge	98,00 m³/s/m
Absolute maximum WL	447,30 m
Length of the structure	322,00 m
Number of openings	9
Opening width	15,00 m
Gate height	18,85 m
Chute width	163,00 m
Chute slope	3 H:1 V
Length of the chute slope	87,50 m
Culverts: 18	5,0 m wide × 7,0 m high
Stilling basin Type II	Froude > 4,5 and velocity > 15 m/s
Length of the stilling basin	95,50 m
Width of the stilling basin	163,00 m
Depth of the stilling basin	37,70 m
Chute blocks	32
End sill blocks	22.

As predicted in the hydraulic model studies, the back currents resulting from the asymmetrical floodgate operations during the diversion and along the spillway operation, caused a lot of solid material to enter the basin brought downstream (380 m³), which caused erosion by abrasion (see Figures 5.62 to 5.72).

After the closure of the 18 culverts, on May 5, 1975, the water was pumped for inspection of the basin. It was found that, in fact, the damages in the basin were extensive, with exposition of reinforced bars, and that there was no time to repair them because filling the reservoir would be quick. (CBDB, 2002). In the first years of operation, it was only possible to monitor the progress of erosion.

In 1980, the basin was emptied and partial repairs were performed. The rocky massif downstream of the basin was consolidated; in addition, small retaining walls were constructed using anchor bars and the depressions were filled with concrete.

The works were completed in 1982. Approximately 3.600 m^3 of concrete and 745 m^3 of mortar were used. In addition, a rigid plan of operation of the gates was established that provided for the operation of the gates with equal openings in order to eliminate the back currents.

The spillway operated intensively over the next 15 years. In the second half of 1997, the basin was again emptied for new repairs that consumed only 85 m^3 of concrete and approximately 160 m^3 of mortar. The stabilized process was considered.

It should be noted that the tests simulating a possible error in the key curve are routine. In these tests artificially lower water levels downstream of the basin are considered to intensify the erosive action of the flow downstream of the hydraulic jump and predict the result in terms of erosion.

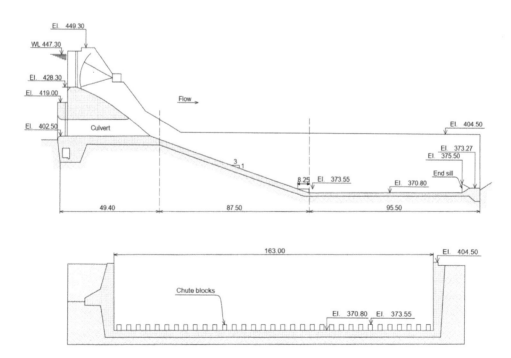

Figure 5.62 Marimbondo HPP. Spillway – Longitudinal profile and cross-section.
Source: (CBDB, 2002).

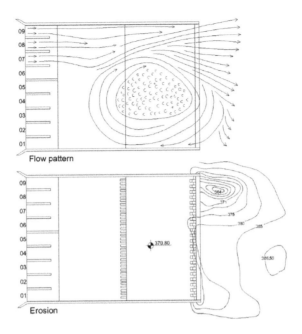

Figure 5.63 Spillway of the HPP Marimbondo. Results of the three-dimensional reduced model tests for Q = 4.000 m³/s, with asymmetrical operation of the gates.

Source: (CBDB, 2002).

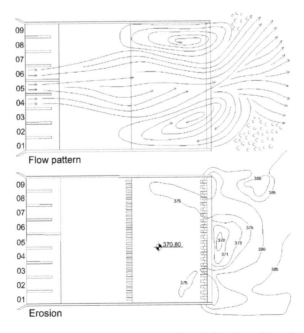

Figure 5.64 Spillway of the HPP Marimbondo. Results of the three-dimensional reduced model tests for Q = 4,000 m³/s, with asymmetrical operation of the gates.

Source: (CBDB, 2002).

Figure 5.65 Spillway of Marimbondo operating during river diversion.

Source: (CBDB, 2002)

Figure 5.66 Spillway of Marimbondo. Solid material deposited in the basin.

Figure 5.67 Damage to the chute after the river diversion.

Figure 5.68 State of concrete reinforcement.

Figure 5.69 State of concrete reinforcement of the stilling basin.

Source: (CBDB, 2002).

Figure 5.70 Spillway of Marimbondo HPP. Reconstruction of the slab of the dissipation basin.
Source: (CBDB, 2002).

This case is classic and, as previously mentioned, in Marimbondo was studied and predicted in the model. The author does not understand why the plan minimizing these operations was not adopted at the beginning of the diversion of the river by the culverts. It is believed that the lesson has been learned. There has been no more record of such cases in Brazil.

Figure 5.71 Spillway of Marimbondo HPP. Reconstruction of the slab of the dissipation basin.
Source: (CBDB, 2002).

Figure 5.72 Spillway of Marimbondo. Rocky massif downstream of the basin after repairs are performed.

Source: (CBDB, 2002).

5.4.3   *Spillway of Nova Ponte*

The Nova Ponte plant was inaugurated in 1987 in the municipality of the same name in the river Araguari (MG). The dam is 141 m high. The surface spillway was positioned on the right abutment, where the rocky massif is a dense basalt, solid and impermeable, as shown in CBDB/MBD II (2000).

Figure 5.73 Nova Ponte HPP.

Source: (CBDB, Highlights, 2006).

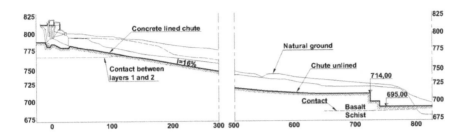

Figure 5.74 Nova Ponte HPP. Longitudinal profile.

Source: CBDB/MBD II (2000).

The chute spillway is 61,60 m wide, with four gates 11 m × 17,35 m, and 699,50 m length; the piers are 3,20 m thick; the discharge capacity is of 6.140 m³/s; the specific discharge is 115 m³/s/m.

The initial 164,50 m of the chute is lined with concrete and covers the contact between basalt I and II layers. The remaining 535 m is unlined and the excavation was done smoothly, without steps, in order not to expose the columnar basalt to erosive process.

This solution, like that of the Serra da Mesa spillway previously exposed, deserves to be highlighted because it provides great savings. The flow is over the unlined rocky massif.

5.4.4   *Spillway of Paulo Afonso IV*

The surface chute spillway of the Paulo Afonso IV HPP (4400 MW) on the São Francisco river, on the left abutment, is founded on a sound granite massif (CBDB/MBD I, 1982).

After the control structure, there is a small concrete slab 76 m long. The rest of the chute is over the rocky massif. The maximum flow capacity is $Q = 10.000$ m³/s. The spillway has eight gates of 11,5 × 19,6 m. The specific discharge is of the order of 90 m³/s/m.

(A)

Figure 5.75 Views of Paulo Afonso IV HPP. A – Spillway closed. B – Spillway operating.

Source: (CBDB, 1982a).

(B)

Figure 5.75 (Continued)

The spillway has been operating since 1979 without any problems. An account of the performance of the structure in the first seven years can be found in Vasconcelos (1986).

5.4.5 Spillway of Crestuma

The construction of the Crestuma HPP (Figure 5.76) on the Douro river, near the city of Oporto (Portugal), 120 MW, ended in 1982, and has aspects that deserve to be highlighted:

• It is a spillway dam, founded in alluvium 40 m thick (sand, gravel and a little silt);

Figure 5.76 Crestuma HPP.

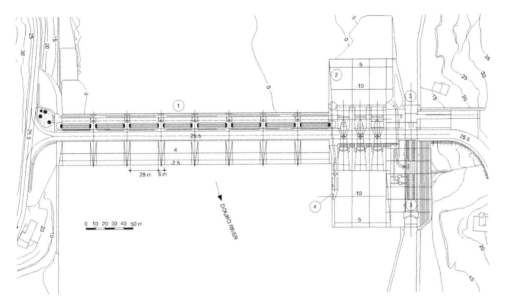

Figure 5.77 Crestuma HPP plant. (1) Dam/spillway. (2) Intake/powerhouse. (3) Navigation lock. (4) Fish ladder.

Source: (Ribeiro, 1973).

Figure 5.78 Hydraulic model, scale 1:80.

Source: (Ribeiro, 1973).

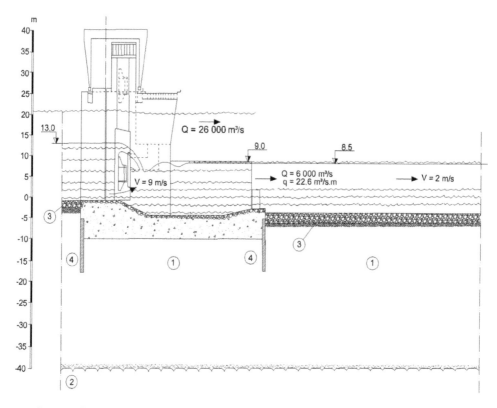

Figure 5.79 Spillway profile. (1) Sand. (2) Rock – shale. (3) Granite blocks. (4) Concrete cutoff.
Source: (Ribeiro, 1976).

• It the time, it was a plant with the largest bulb turbines in the world (6.8 m in diameter).

The plant is the last of the river Douro. It drains practically the entire area of the basin. The general arrangement of the work comprises a dam/spillway, intake/powerhouse, navigation lock and fish ladder.

The spillway design discharge is 26.000 m³/s. The dam has eight double-leaf vertical lift gates or caterpillar gates (Figures 5.76 and 9.27), 28 m wide, 13,7 m high.

A summary of the project solution is presented in Figures 5.77 to 5.82, taken from Ribeiro *et al.* (1967, 1973, 1976, 1979).

The layer of protective granular material downstream of the spillway was studied in a two-dimensional reduced model on the 1:36 scale (Ribeiro, 1973). It was determined that this layer should be 70 m long and be constituted as follows: the initial stretch, 30 m after the spillway, should be 4 m thick and 50% of the blocks with 5,4 *t* (*D* = 1,50 m); afterwards, the layer should be 2,50 m thick with blocks of 1,0 *t* (*D* = 0,90 m).

Figure 5.80 Two-dimensional reduced model, scale 1:36. Testing of granular protective material.
Source: (Ribeiro, 1973).

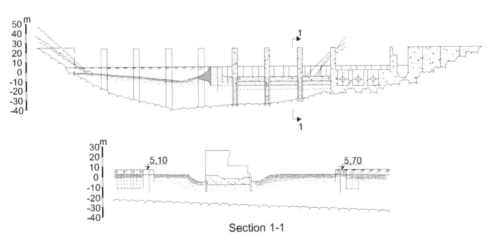

Section 1-1

Figure 5.81 Crestuma HPP. Longitudinal cut and section during construction
Source: (Ribeiro, 1976).

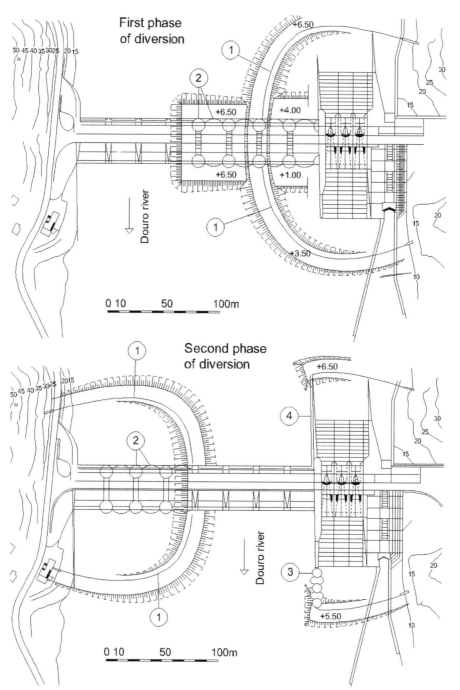

(1) Non-reinforced diaphragm wall cofferdam; (2) Reinforced diaphragm wall cofferdam; (3) Sheet piles cells; (4) Powerhouse cofferdam.

Figure 5.82 Crestuma HPP. River diversion.
Source: (Ribeiro, 1979).

5.4.6 Spillway of Lower Notch

The hydroelectric dam of Lower Notch (228 MW), owned and operate by The Hydro-Electric Power Commission of Ontario, was completed in 1971 and has been in successful operation since that time (Atkinson and Overbeeke 1973).

The project is located near the mouth of the Montreal river in the province of Ontario, approximately 400 km north of the city of Toronto. The Montreal river is a tributary of the Ottawa river and discharges into lake Timiskaming, a widened portion of the Ottawa river. The project layout is shown in the Figure 5.83.

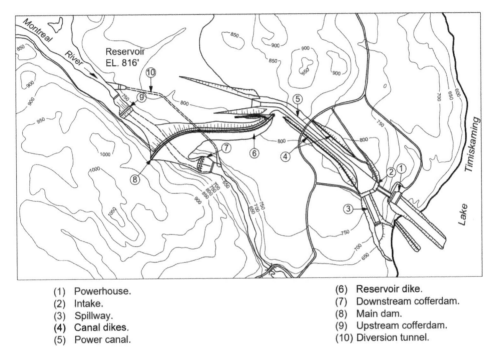

(1) Powerhouse.	(6) Reservoir dike.
(2) Intake.	(7) Downstream cofferdam.
(3) Spillway.	(8) Main dam.
(4) Canal dikes.	(9) Upstream cofferdam.
(5) Power canal.	(10) Diversion tunnel.

Figure 5.83 Lower Notch HPP. Montreal river, Ontario province, Canada.

Source: (Atkinson and Overbeeke, 1973).

The bedrock at the site is consists primarily of competent graywacke, graywacke conglomerate, argilites and arkose sandstone of the Cobalt series, which are of Precambrian age. The regional terrain is characterized by a presence of pronounced linear depressions, oriented generally in a northwesterly direction, in which lake Timiskaming and the Montreal river lie. A similar set of linear depressions trends northeast but is less pronounced. A third set of joints, which consists of the bedding planes, strikes to the northwest and dips rather flatly to the southwest. While the original bedding planes are difficult to distinguish in some areas, they are sometimes marked by a filter material of broken rock, clay and slaty bands. Although the joints corresponding to the bedding planes are generally widely spaced, a zone of closely spaced bedding planes was found about halfway down the penstock excavation. The rock in the area of the spillway flip bucket and the area immediately downstream of the flip bucket comprises the best quality rock found at the site, with joint spacing from 0,30 m to 1,50 m or more.

Figure 5.84 Lower Notch HPP. From lake Timiskaming.
Source: (Atkinson and Overbeeke, 1973).

Figure 5.85 Lower Notch HPP. Operation of spillway gate 3. ($Q = 322$ m^3/s).
Source: (Atkinson and Overbeeke, 1973).

Figure 5.86 Lower Notch HPP. Operation of spillway gates 1 and 3 ($Q = 706$ m^3/s).
Source: (Atkinson and Overbeeke, 1973).

Figure 5.87 Lower Notch HPP. Jet landing area.
Source: (Atkinson and Overbeeke, 1973).

Figure 5.88 Lower Notch HPP. Jet landing area.
Source: (Atkinson and Overbeeke, 1973).

The gross head is 70 m, approximately, and the spillway capacity is of 1.440 m^3/s (TR = 1:10.000 years). The Figures 5.84 to 5.88 show the spillway running. Erosion with 8,0 m depth occurred as planned in the tests without problems for the safety of the plant.

5.4.7   Spillway of Sardar Sarovar

The Sardar Sarovar plant (Figures 5.89 and 5.90) was built on the Narmada river, which flows into the gulf of Khambhat, Gujarat, India. It was inaugurated in September 2017 and has two powerhouses: the main with 1.200 MW and the channel powerhouse with 250 MW. The dam is 163 m high and 1.210 m long.

It has a large spillway, with a total discharge capacity of 85.000 m^3/s. It is interesting to note that it is separated in two sectors:

Figure 5.89 Sardar Sarovar spillways.

Figure 5.90 Sardar Sarovar. Spillways operating.

- the main, in hydraulic jump, has 23 gates of 18,30 m × 18,00 m;
- the auxiliary, in ski jump, has seven gates of 18,30 m × 16,75 m.

5.4.8 Spillways of Salto Grande

Salto Grande HPP, 74 MW, was built on the Paranapanema river, a tributary of Paraná river, between São Paulo and Paraná States, in the southern region of Brazil (Figures 5.91 to 5.94). It was inaugurated in April 1958. The dam is 22 m high. The local geology is formed by sound basaltic layers of the Serra Geral Formation, with the presence of sandstone dikes. The surveys conducted in the spillway region revealed a little to very fractured vesicular basalt, with the fractures filled with calcite.

Figure 5.91 Salto Grande. Powerhouse, buttress dam and spillways.

The plant has two spillways, Creager profile with roller bucket dissipator, with a total discharge capacity of 8.500 m³/s:

- the main with eight spans, L = 12,5 × H = 11,6 m, and Q = 7.910 m³/s;
- and the emergency with 14 spans, L = 11,5 × H = 1,5 m, and Q = 590 m³/s.

Figure 5.92 Salto Grande. Downstream view of the main spillway.

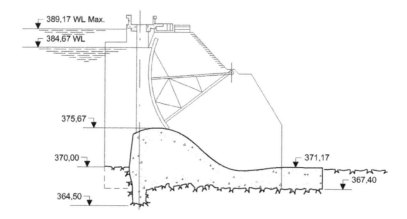

Figure 5.93 Salto Grande. Main spillway section.

Figure 5.94 Salto Grande. Small craters of erosion.

According to Magela and Rodrigues (1991), the downstream erosion was minimal in the form of small holes in the basaltic massif, as shown in Figure 5.94 – an actual picture. In all the hydropower plants built by CESP in basalt region, it was verified that this type of rock has good resistance to erosion.

Chapter 6

Pressure forces downstream of dissipators

Hydrodynamics pressures caused by water jets downstream of spillway energy dissipators can be estimated in two ways: through hydraulic calculations or through measurements in hydraulic models. They were calculated for the ski jump of the Tucuruí spillway in 1981. The maximum pressure fluctuation reaching the bottom of the river downstream was of 13,0 t/m^2 (see Figure 6.4). After more than 35 years of operation, there was no occurrence of any erosive process.

Magela and Brito presented in 1991 the work "The Use of a Skijump Spillway in Heavily Fractured Rock Masses – the Jaguara HPP Case" containing the methodology to calculate the efforts (Figures 6.7 to 6.9). At the beginning of the operation, the jet impacted directly on the sound rock and the pressures ranged from 3,0 to 4,0 t/m^2 (Figure 6.8). Due to the characteristics of the massif, with very fractured quartzite, erosion developed rapidly (Figures 5.4 to 5.10).

With respect to hydraulic aspects, Magela and Brito (1991) presented a summary of the subject in the paper "Scour in Launch Basins, Hydraulic and Geotechnical Aspects," published in *CBGB Magazine*, N. 03/96 (1996). One should refer to the work of Sobrinho and Infanti (1986), "Erosion of Rock Masses Subject to Flow Action, Some Geomechanical and Hydraulic Aspects," and the various works of Bollaert (2006, 2010, 2012), among others.

6.1 Hydrodynamics pressures downstream of ski jumps

The methodology for estimating the hydrodynamics pressures downstream of a ski jump is based on the theory of free jet turbulence developed and tested by Prandtl for air – see Hartung and Haüsler (1973) and Haüsler (1983). The behavior is analogous to that of a jet of water coming from a nozzle and plunging into an unlimited water depth.

The pressures obtained were higher than the real ones, therefore, in favor of safety, since the beneficial influence of the aeration in the dissipation of part of the energy in the aerial phase was neglected, given the complexity of the phenomenon.

Haüsler also observed a similar behavior in the studies in model of free fall jets diving in a mass of water. When the jet enters an unlimited space filled with the same fluid, it disperses almost completely linearly (Figures 6.1 and 6.2). After the jet dips into the body of water, the "jet core" of length "Yk" is formed downstream. Its boundaries are also linear, as shown in Figure 6.2.

Along the path, from the point of impact to Yk, the zone is characterized by the fact that the velocity is constant and the pressure distribution is hydrostatic. Measured data show

Figure 6.1 Circular jet linear dispersion (15 cm in diameter). Katse dam, Lesotho.
Source: (Hartung and Hausler, 1973); (ae.americananthro.org).

that, in planes perpendicular to the jet, the velocity and hydrodynamic pressure distribution obeys a Gaussian curve (Normal Distribution).

Under the hydrostatic pressure condition, the sum of the external forces is equal to zero along the entire jet path. Therefore, the amount of movement of the jet along its trajectory is equal to that which enters the water mass.

Using the fundamental equations of hydrodynamics, momentum/impulse, continuity and energy, important results are obtained for the hydraulic project engineer, as shown next.

$$\frac{J}{Ju} = \frac{\int_0^\infty V^2 dS}{Vu^2 Su} = 1 \tag{6.1}$$

$$\frac{Q}{Qu} = \frac{\int_0^\infty Q dS}{VuSu} > 1 \tag{6.2}$$

$$\frac{E}{Eu} = \frac{\int_0^\infty V^3 dS}{Vu^3 Su} < 1 \tag{6.3}$$

The core length is defined using the tangent of the internal diffusion angle:

$$tg\alpha_i = \frac{Bu}{Yk} \text{ or } \frac{Ru}{Yk} \tag{6.4}$$

Several researchers have shown that the internal diffusion angle varies between 4° and 6°, depending on the Reynolds number. For the problems of applied hydraulics, the aforementioned authors adopt, $tg\alpha_i = 0{,}10$ for both rectangular and circular jets.

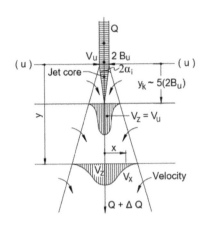

According to Hartung and Häusler (1973)

$tg\ \alpha_i = Bu/Yk;$

As $4° < \alpha_i < 6°$ on practice it is adopted $tg\ \alpha_i \simeq 0{,}10$

a) <u>Unlimited tailwater</u>

$$F_{exterior} = 0\ !$$

Rectangular jet:

$$J = 2 \cdot \rho \cdot L \cdot \int_{x=0}^{x=\infty} V_x^2 \cdot dx = const. = j_u$$

For $y > y_k$:

$$V_x = V_u \left(\frac{y_k}{y}\right)^{\frac{1}{2}} \cdot e^{-\frac{\pi}{8}\left(\frac{x}{Bu} \cdot \frac{y_k}{y}\right)^2}$$

Velocity distribution

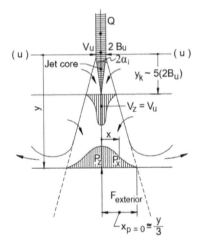

b) <u>Limited tailwater</u>

$$F_{exterior} \neq 0\ !$$

Rectangular jet:

$$F = 2 \cdot L \cdot \int_{x=0}^{x=\infty} P_x \cdot dx = const. = j_u$$

For $y > y_k$:

$$P_x = \rho \cdot g \cdot \frac{V_u^2}{2g} \cdot \frac{y_k}{y} \cdot e^{-\frac{\pi}{16}\left(\frac{x}{Bu} \cdot \frac{y_k}{y}\right)^2}$$

Pressure distribution

Figure 6.2 Free jet turbulence theory.

Source: (Hartung and Hausler, 1973).

6.1.1 *Important results for the project engineer*

6.1.1.1 *Thickness of water height to protect the rocky massif*

From the previous equation we obtain:

$$Yk \sim Bu/0,10 \sim 10Bu \sim 5(2Bu) \text{or} \sim (2Ru) \tag{6.5}$$

or even:

$$Yk/Bu \sim 10 \tag{6.6}$$

where

2*Bu* and 2*Ru* represent the width and diameter of the jet on the impact surface.

This implies that with a jet 1,0 m in diameter, a water height of 5,0 m does not reduce the jet velocity and the dynamic pressure of the jet core in the ground.

It should be noted that the width, or diameter, of the jet impact zone is difficult to assess, given the effects of aeration and jet divergence.

In order to estimate the pressure on the impact of the jet, $2Bu = q/Vu$ is adopted, neglecting the effects quoted and thus favoring safety.

6.1.1.2   Dissipation of energy along the nucleus of the jet

Along the length of the *Yk* jet core, only 20% of the energy is dissipated for rectangular jets and 30% for circular jets.

6.1.1.3   Zero pressures (end of jet influence)

Studies of Cola (1965) have shown that the end of the influence of rectangular jet pressures occurs when $Y = 40\ (2\ Bu)$. Hartung and Haüsler (1973) have shown that the end of influence of circular jet pressures occurs when $Y = 20\ (2\ Bu)$.

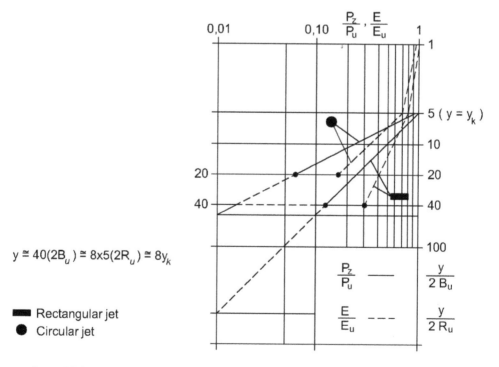

$y \cong 40(2B_u) \cong 8\times5(2R_u) \cong 8y_k$

■ Rectangular jet
● Circular jet

Figure 6.3 Theoretical model of jet behavior.

Source: (Hartung and Haüsler, 1973).

6.1.1.4 *Hydrodynamics pressures*

According to these authors, the hydrodynamics pressures in the zone of impact of the jet up to the depth $Y = Yk$ along the center line of the jet is constant and equal to:

$$Pu = \frac{Vu^2}{2g} \tag{6.7}$$

For depths $Y > Yk$, the pressures are calculated by the expression:

$$\frac{P}{Pu} = \frac{Yk}{Y} e^{-\frac{\pi}{16}\left(\frac{Yk}{Bu}\frac{X}{Y}\right)^2} \tag{6.8}$$

which can be written as

$$\frac{P}{Pu} = \frac{Yk}{Y} e^{-\frac{\pi}{16}\left(10\frac{X}{Y}\right)^2} \tag{6.9}$$

The bulb and the hydrodynamics pressures diagram are calculated by solving Equation 6.8. After the substitution of the known parameters, Pu and Yk, values of "P" and "Y" are determined by "X" values. With the pairs (X, Y), it is possible to construct the pressure diagrams.

6.1.1.5 *Impact angle of the jet on the downstream water surface*

$$tg\alpha_u = \frac{1}{\cos\alpha}\sqrt{sen^2\alpha + \frac{H1}{H0}} \tag{6.10}$$

6.2   Example: pressure bulb in the pre-excavation downstream of the Tucurui spillway

Q	= discharge (TR=10.000 years)	100.000	(m³/s)
WLr	= reservoir water level	74,00	(m)
WLd	= downstream water level	24,50	(m)
H	= gross head = $WLr - WLd$	49,50	(m)
B	= apron lip width	552,00	(m)
q	= specific discharge at apron lip	181,00	(m³/s/m)
L	= length of Creager profile	90,00	(m)
hf	= head loss in the Creager profile	0,50	(m)
H_0	= head at bucket lip (WLr – lip elev. – head loss until the lip)	41,00	(m)
H_1	= vertical drop from the lip to tailwater surface	8,00	(m)
V_L	= takeoff velocity: $V_L = \sqrt{2gH_o}$	 28,40	(m/s)
d_L	= water depth in the lip of the jet takeoff:		
d_L	= q/V_L (m);	6,00	(m)
Vu	= velocity of the jet when impacting the WLd or on the rock top:		
Vu	= $\sqrt{2gH}$	31,00	(m/s)
α	= the bucket takeoff angle	32°	

L_T = throw distance (HDC 112–8)

L_T = $H_o sen\ 2\theta + 2H_o cos\theta \sqrt{sen^2\theta + {}^Y/_{H_v}}$

L_E = effective range of the jet calculated as presented by Martins (1977)

$2Bu$ = the width of the jet in the impact zone

$2Bu$ = $q/Vu = 181/31,00$ 5,84 (m)

Yk = jet core length

Yk = $Bu/0,10$ 29,20 (m)

Pu = hydrodynamic pressure on jet impact

Pu = $Vu^2/2g$ 43,50 (t/m^2)

α_I = jet impact angle

$tan\alpha_I = \frac{1}{cos\alpha}\sqrt{sin^2\alpha + \frac{H1}{Ho}}$ 0,81

α_i 39°

Solving the equation 6.8 we have:

$$X = \sqrt{-0,05Y^2(lnPY - lnPuYk)}$$

Substituting the values of "Pu" and "Yk" for each discharge and arbitrating values for "Y" and "P," we have the values of X (shown in the Table 6.1).

Table 6.1 Pressure bulb for Q = 100.000 m^3/s

| P (t/m^2) | 40 | 35 | 30 | 25 | 20 | 15 | 10 | 5 | 1 |
Y (m)	X (m)								
30	1,64	2,95	3,96	4,88	5,82	6,84	8,07	9,81	12,99
40		2,19	4,40	6,10	7,76	9,63	12,17	16,64	
50			2,64	5,50	8,14	10,81	14,27	20,12	
60				3,28	7,91	11,64	16,13	23,45	
70					6,88	12,11	17,79	26,66	
80					4,38	12,20	19,25	29,76	
90						11,87	20,53	32,76	
100						11,01	21,63	35,67	
110						9,43	22,54	38,5	
120						6,57	23,29	41,24	
130							23,85	43,92	
140							24,23	46,52	
150							24,42	49,06	
160							24,41	51,53	

Figure 6.4 shows the graph of the pressure bulb. It is observed that the maximum pressure fluctuation (P) reaching the rock mass is of the order of 13 t/m^2.

The total hydrodynamic pressure is equal to the sum of the mean hydrostatic pressure, with the pressure fluctuation: $Pt = (64,50 \pm 13)$ t/m^2

Figure 6.5 shows the position of the pressure cells installed in the pre-excavation model downstream of the Tucuruí HPP spillway built at FCTH-USP. The experimental data measured in the reduced model show for the pressure cell T2: $Pmax = 81$ t/m^2; $Pmed = 66$ t/m^2 and $Pmin = 57$ t/m^2, demonstrating good tuning with the calculated data.

Figure 6.6 shows the variation of the coefficient of turbulence.

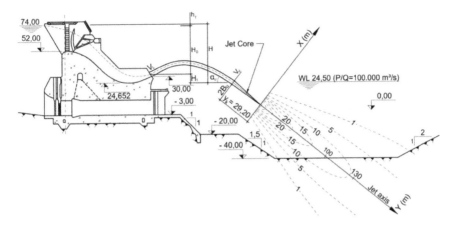

Figure 6.4 Tucuruí HPP Spillway. Pre-excavation pressure bulb. (Pre-excavation in the reduced model and in the work – Figures 10.19 and 10.20).

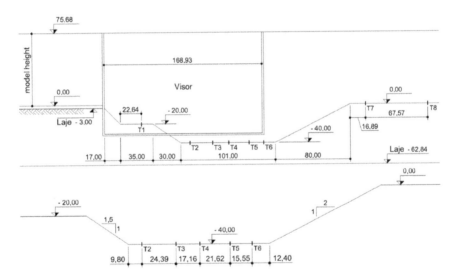

Figure 6.5 Model of the Tucuruí HPP spillway. Pressure sockets.

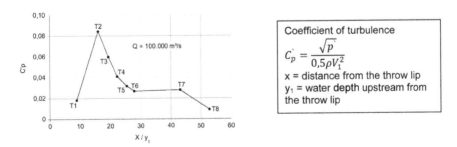

Figure 6.6 Model of the Tucuruí HPP spillway. Variation of the coefficient of turbulence.

6.3   Example: pressure bulb in the pre-excavation downstream of the Jaguara spillway

Magela and Brito presented an analysis of this case in 1991, and the results are presented in Figure 6.8. The spillway cross-section is shown in Figure 5.12. Several photos were previously presented in section 5.2.

Figure 6.7 Jaguara HPP. Observation: the actual situation of this plunge pool is presented in Figure 5.9.

Source: Magela and Brito (1991).

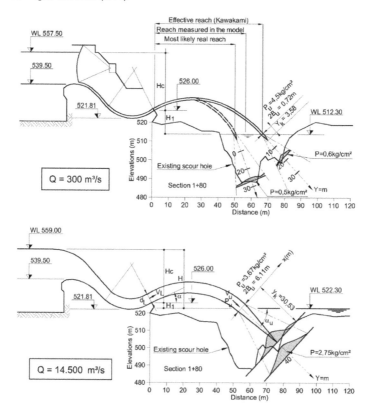

Figure 6.8 Jaguara spillway. Pressure bulb in the plunge pool.

Source: (Magela and Brito, 1991).

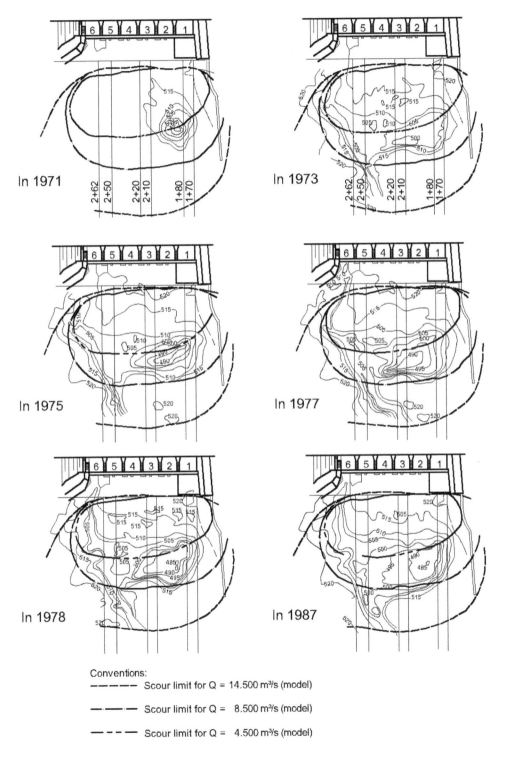

Figure 6.9 Scour downstream of the Jaguara spillway over time.

Source: (Magela and Brito, 1991).

6.4   Hydrodynamics pressures downstream of stilling basins

A flow parallel to a rocky surface as it occurs at the outlet of a dissipation basin will transmit to the rocky mass hydrodynamics pressures; the larger, the greater the flow velocities. The velocity estimation should be done as explained in section 5.3. To evaluate the downstream pressures of the stilling basins, the methodology proposed by Reinius (1986) is used.

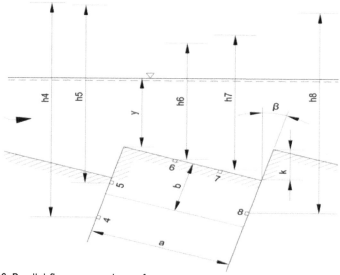

Figure 6.10 Parallel flow to a rocky surface.

*Source:*Reinius (1986). (Magela and Brito, 1996).

These pressures can be estimated using the classical expression:

$$P = \rho C \frac{V^2}{2} = 0,5CV^2 \tag{6.11}$$

where:
ρ = density of water = 1,0 t/m^3;
V = flow velocity (m/s);
C = coefficient of pressure that varies according to the angle of the fractures (Figure 6.11).

The efforts acting on the different faces of a potentially removable block are also illustrated in Figure 6.11. Turbulent pressure fluctuations can be estimated by the expression:

$$P' = \pm 3\tau = \pm \rho f \frac{V^2}{8} = \quad \pm 0,4 f V^2 \tag{6.12}$$

where:
τ = mean shear stress at water-rock contact;
f = coefficient of friction ranging from 0.03 to 0.08 for an excavated rock surface (Reinius, 1970, 1986).

The total hydrodynamic pressure will be given by the sum of the two parcels:

$$Pt = 0,5CV^2 \pm 0,4 f v^2 (t/m^2) \tag{6.13}$$

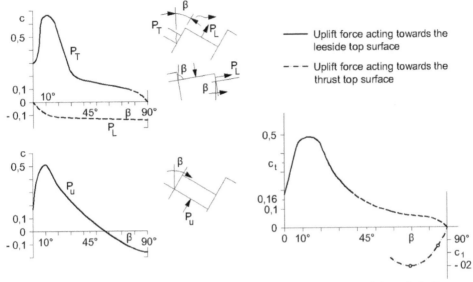

Greatest pressure coefficients C of forces P_T on the thrust surface, P_L on the leeside surface, and P_u in a joint parallel with the leeside surface.

Greatest pressure coefficients C_t for the total uplift force acting on a block of rock.

Figure 6.11 Coefficients of pressure.

Source: (Reinius 1986; Magela and Brito, 1996).

6.5   Hydrodynamics pressures downstream Canoas I HPP spillway stilling basin

The following is an example of the hydrodynamics pressures downstream of the watershed dissipation basin at the Canoas I HPP (Figures 6.12A and B).

In the estimation of the hydrodynamics pressures, $Pt = 0,5CV^2 \pm 0,4fV^2$, $C = 0,5$ and f varying between 0,03 and 0,08 were adopted.

Figure 6.12A Canoas I HPP.

Source: (www.duke-energy.com.br).

Figure 6.12A (Continued)

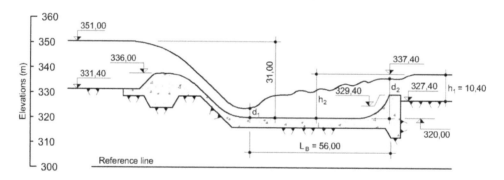

Figure 6.12B Longitudinal section of the spillway and dissipation basin of the Canoas I HPP.

Q	= discharge (TR = 1.000 years) 5.000 (m^3/s)	
WLr	= reservoir water level 351,00 (m)	
WLd	= downstream water level 337,40 (m)	
H	= gross head = WLr − WLd 13,60 (m)	
B	= apron lip width 70,50 (m)	
q	= specific discharge at apron lip 71,00 $(m^3/s/s)$	
CF	= basin elevation 320,00 (m)i = head on the basin 31,00 (m)	
H_d	= dimensioning head of the spillway 13,50 (m)	
V_1	= velocity at basin entrance	

$$V_1 = 097\sqrt{2g\left(Z - \tfrac{H_d}{2}\right)}\ 21,16\ (m/s)$$

d_1	= upstream height of jump: $d_1 = q/V_1$ 3,34 (m)	
F_1	= Froude number at basin entrance: $F_1 = V_1/(gd_1)^{0.5}$ 3,71 < 4,5	

$$d_2 = 05d1\sqrt{1 + 8F_1^2} - 1\ 16,00\ (m)$$

d_2	= downstream height of jump:	
h_2	= height at basin exit 17,40 (m)	
hj	= height downstream of the basin 10,40 (m)	
h_2/d_2	= degree of drowning (submergence) 10,00 (%)	
L_B	= length of basin: $L_B = 0,6 \times 6 \times d2$ 56,00 (m)	
V_j	= velocity downstream the end sill: $V_j = q*/hj$ 7,10 (m/s)	

* Valid consideration because there is in this case no lateral expansion of the flow immediately downstream of the sill; the velocities measured in the model varied between 7 and 9 m/s.

From the results presented here it can be seen that the pressures, in this case, vary very little as a function of the roughness.

Figure 6.12B Longitudinal section of the spillway and dissipation basin of the Canoas I HPP.

f	0,03	0,04	0,05	0,06	0,07	0,08
$Pt(t/m^2)$	13,21	13,41	13,64	13,81	14,01	14,22

The flow pattern can be seen in Figure 6,13, the Balbina spillway. The typical cut is shown in Figure 6.14.

Figure 6.13 Balbina HPP. Spillway with stilling basin – hydraulic jump. Observe the back currents attacking the downstream slope of the dam.

Source: (Magazine Eletronorte 25 years, 1998).

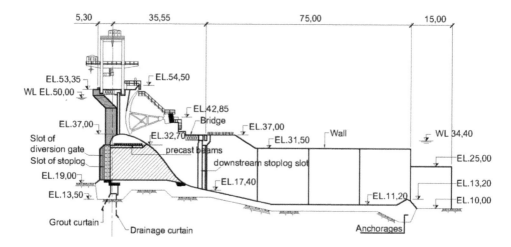

Figure 6.14 Balbina HPP – Spillway and stilling basin; four radial gates 13,50 m × 15,30 m. Q = 5.840 m^3/s.

Source: (MBD, 2000).

Chapter 7

Evaluation of the scour

The prediction of the scour pit is still very imprecise, due to the large number of factors that intervene in the phenomenon: the incident jet form; the energy of the jet/gross head; the specific discharge; the degree of aeration of the jet; the height of water downstream; the rocky matrix of the riverbed and its degree of homogeneity; the degree of alteration and diaclasing of the rock and the possible existence of geological faults; the frequency of operation of the spillway; and the frequency of asymmetric sluice operations.

The photographs in Figure 7.1 were taken by the author during a visit to the Jaguara plant to study this scour case (Magela and Brito, 1991). Other photos of Jaguara are presented in sections 5.2 and 6.3.

Figure 7.1 Jaguara spillway: scour pit and jet throw distance.

The jet influences erosion through its velocity, throw distance and shape on impact, conditioning the location of the excavation and the geometry of erosion. The behavior of the jets, according to Martins (1973), depends on their initial velocity and turbulence, their characteristic size at the board lip, diameter or height, the shape and path factors in the air. Due to the resistance of the air and the consequent aeration, the effective throw distance of the jet is inferior to the obtained theoretically.

Emulsification of water is beneficial for the reduction of its scour power. This emulsification of water is beneficial for the reduction of its erosive power. Currently, this influence is already considered in the estimation of the scour (Mason, 1984a). The greater the height of the water downstream, the less the residual energy with which the jet will affect the rock mass and, consequently, the lower the erosion.

The increase in the specific discharge, caused by the advance in the technology of drive and operation of gates and that has led to increasing the size of gates, tends to aggravate the risk of scour (Neidert, 1980a). The evolution of the erosion process is related to the frequency of operation of the spillway, being faster the higher the spillway operating frequency, depending on the magnitude of the discharges spilled. The asymmetric operation of gates facilitates the intensification of the back currents, accentuating the erosive power of the flow.

In the evaluation of downstream erosions, reduced models and traditional formulas deduced from data obtained from model and prototype tests are used.

Regarding the hydraulic-geological/geotechnical interface of the issue, besides the projects carried out in the USSR in the 1960s, there were few attempts until the 1980s to understand the dynamic interaction of hydraulics with geology in the design (Sobrinho and Infanti, 1986).

The evolution of the theme necessarily passes through this understanding and through the identification and estimation of these efforts. Pressure fluctuation measurements and their association with rock mass resistance to hydrodynamic stresses may be another way of predicting erosion. See more recent studies cited in section 6.3, using the mathematical model CSM – Comprehensive Scour Model (Bollaert, 2004a, 2006, 2010, 2012). See also Noret (2012). It is worth highlighting the several works developed at IPH-UFRGS guided by Professor M. G. Marques, many of them available on the internet.

Whatever the method of prediction, the difficulty of estimating erosion most accurately is because it is difficult to assess the strength of the rock mass.

7.1　Scour holes in hydraulic models

As seen in Chapter 3, evaluations of erosive processes are performed in reduced models, in which the susceptible area of the whole process is represented and two types of tests are performed: one with a loose moving bottom and one with a mobile bottom with some degree of cohesion. This technique only leads to qualitative estimates of the evolution tendency of the erosion pit.

In section 10.5 are presented the studies realized for the Tucuruí HPP. In the studies with a loose moving bottom, without cohesion, in the modeling of the area that could be subject to the erosive process, grave is used to represent the potentially removable blocks of the rocky massif (scale of the model). This type of test applies when, in the region in focus, the mass is very fractured or of low resistance to dynamic efforts. In these tests, a good indication of the limit of the erosion crater is obtained.

In the tests with a cohesive mobile bottom, the resistance of the massif is simulated using, for example, aluminum cement or gypsum. The definition of the mortar trace is complex and is not part of the scope of this book. It is recommended to consult the works of Feldman (1970), about the project of Grand Rapids, and of Lencastre (1982) on the project of Picote.

The knowledge of the erosions in the prototypes, their association with the operating conditions that caused them and the geological survey of the area susceptible to the erosive process is valuable information to assess the quality of the elaborated project and the evolution in the theme, as well as to enable a better plan and guide future project studies.

7.2   Estimate of scour holes depths; Veronese equation

Several formulas were developed for the estimation of depth of erosion, based on experimental prototype studies. Mason and Arumugan (1985) performed a detailed critical analysis of these formulas as shown in Table 7.1.

Group I formulas have the general form:

$$D \cong K\frac{q^{x}H^{y}}{d^{z}} \tag{7.1}$$

where:
K, x, y and z are constant for each of the formulas (see Frame 7.2);
D = depth of erosion;
q = specific discharge;
H = total head;
d = diameter of the characteristic block.

An exception is the formula of Bizaz and Tschopp, which has the expression:

$$D = Kq^{x}H^{y} - Kd \tag{7.2}$$

The authors of Group I, according to Mason and Arumugan (1985), seem to have derived their formulas from experiments in which the parameters were modified in order to translate the observed erosive effect. This fact led to an "x" coefficient close to 0,6 in all of them and to a "not so constant" value in the range of 0,2 to 0,3. There is, however, a greater variation in the value of "z," which is between 0 and 0,5, and also in the characteristic diameter to be used.

Most authors preferred d_{90}, assuming that the smaller particles will be rapidly eroded and that the final depth of the pit will be defined by the larger particles in the riverbed.

Veronese (1937) analyzed in his experiments the erosion exerted by a jet in free fall on the material without cohesion, as pebbles, since most of the valleys of Italy, for which its formula applied, has flood alluvial. He recommended that the average diameter of the material be obtained from samples of the riverbed, or valley, under study. The formula Veronese B (see Frames 7.1 and 7.2) arose from the observation in tests that, from a certain diameter, erosion became independent of this factor.

Frame 7.1 Formulas for estimating the depth of the pit.

Group	Author	Year
I	Schoklitsch	1932
Veronese A	1937	
Veronese B	1937	
Petrasaev	1937	
Eggenburger	1944	
Hartung	1959	
Franke	1960	
Damle A	1966	
Damle B	1966	
Damle C	1966	
Chee and Padivar	1969	
Bizas and Tschopp	1972	
Wu	1973	
Chee and Kung	1974	
Martins B	1975	
Taraimovich	1978	
Machado	1980	
Sofrelec	1980	
Incyth	1981	
Pinto	1983	
II	Jaeger	1939
Martins A	1973	
III	Lencastre	1961
Cola	1965	
David and Sorensen	1969	
Hartung and Häuler	1973	
IV	Mikhalev	1960
Rubinstein	1965	
Solovyeva	1965	
Yuditskii	1965	
Mirtskhulava A	1967	
Mirtskhulava B	1967	
Mirtskhulava C	1967	
Svorkin et al	1975	
V	Thomas	1953

According to Pinto (1983), the results of the prototypes of São Simão HPP and Foz do Areia HPP seemed to suggest an expression (Equation 7.3) valid for basalts.

$$D = 0,60 q^{0,54} H^{0,225} \tag{7.3}$$

For details on the formulas of the other groups, it is recommended to consult Mason and Arumugan (1985). Based on their analysis, these authors concluded that the formulas that best applied to reality were those of Group I and proposed the following formula (Equation 7.4) for an estimate of the pit, based on model and prototype data, with

Frame 7.2 Coefficients for use in equation 7.1.

Author	K	x	y	z	d
Schoklitsch	0,521	0,57	0,2000	0,3200	d_{90}
Veronese A	0,202	054	0,2250	0,4200	d_{50}
Veronese B	1,900	0,54	0,2250	0,0000	-
Petrasaev	3,900	0,50	0,2500	0,2500	d_{50}
Eggenburger	1,440	0,60	0,5000	0,4000	d_{90}
Hartung	1,400	0,64	0,3600	0,3200	d_{85}
Franke	1,130	0,67	0,5000	0,5000	d_{90}
Damle A	0,652	0,50	0,5000	0,0000	-
Damle B	0,543	0,50	0,5000	0,0000	-
Damle C	0,362	0,50	0,5000	0,0000	-
Chee and Padivar	2,126	0,67	0,1800	0,0630	d_{50}
Bizas and Tschopp (Eq. 8.2)	2,760	0,50	0,2500	1,0000	d_{90}
Wu	1,180	0,51	0,2350	0,0000	-
Chee and Kung	1,663	0,60	0,2000	0,1000	d_{50}
Martins B	1,500	0,60	0,1000	0,0000	-
Taraimovich	0,663	0,67	0,2500	0,0000	
Machado	1,350	0,50	0,3145	0,0645	d_{90}
Sofrelec	2,300	0,60	0,1000	0,0000	-
Incyth	1,413	0,50	0,2500	0,0000	-
Pinto	1,200	0,54	0,2250	0,0000	-

exponents and constant variables.

$$D \cong K \frac{q^{x} H^{y} h^{w}}{g^{v} d^{z}} \tag{7.4}$$

where:
$x = 0,60 - H/300$; $y = 0,05 - H/200$; $w = 0,15$; $v = 0,30$; $z = 0,10$,
D = depth of erosion;
$K = 6,42 - 3,10 \, H^{0,10}$;
q = specific discharge;
H = total head;
h = downstream height of the water;
g = acceleration of gravity;
d = diameter of the characteristic block.

Complementing this, the erosion pits measured in several prototypes are presented in Figure 7.2, comparing them with the pits estimated by formulas of different authors. Note that the Veronese equation is the envelope.

This illustration (Figure 7.2) was used to try to define the values of the coefficient K as a function of the characteristics of the rock in the jet dive basin, as described by Magela and Brito (1996).

$K = 0,525$ for very hard rocks (Foz do Areia, São Simão)
$K = 1,2$ for medium rocks
$K = 1,4$ for less resistant rocks (Tarbela), alluviums (Colbun), or rock resistant with long exposure to large flows (Itaipu)
$K = 1,9$ boundary value, Veronese envelope – very fractured, erodible rocks (Jaguara).

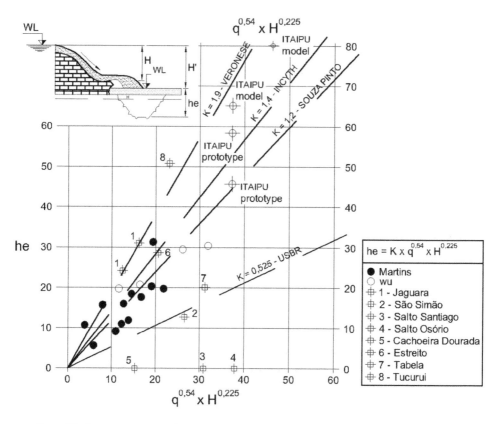

Figure 7.2 Erosion measured downstream of several spillways × erosion estimates using multiple authors' formulas.

Source: (CBDB, 2002)

Table 7.1 Erosion of spillways in ski jump in Russia (Taraimovich, 1978)

Dam (Year)	River	Spillway data			Rock	Maximum scour	
		Q (m³/s)	H (m)	P (m)		L (m)	P (m)
Deproges (32)	Dnieper	25.150	35	17	Gneiss	50	8
Farkhad (48)	Syr Darya	4.520	14	5	Sandstone	75	18
Bukhtarma (60)	Irtish	44.520	67	14	Gabbro	Nd	-
Irikla (58)	Ural	7.805	23	18	Fractured tuff	118 L	-
Bratsk (64)	Angara	17.500	95	12	Diabasio	65 L	12
Ust-Ilim (80)	Angara	14.900	80	11	Diabasio	73 L	12
S-Shu (80)	Yenisei	15.900	166	18	Schist	132 L	41
Krasnoyarsk (72)	Yenisei	12.000	80	18	Granite	102	17
Inguri (85)	Inguri	2.500	225	34	Limestone fract.	124	46
Toktogol (78)	Naryn	3.400	183	20–30	Limestone fract.	Nd	5–10

H = total head (m); P = depth of erosion from the rocky top; L = laboratory.

Sobrinho and Infanti (1986) stated that the understanding of rock mass performance in the prototype is the best criterion for the design of a new structure. The good sense of engineering and the experience of the engineer count more than an empirical formula or a theoretical study. See also Akmedov (1968).

Erosion data downstream of spillways with ski jumps in Russia (Taraimovich, 1978) are presented in Table 7.1.

7.3   Yuditskii iterative method

The following text is a summary of the translation of the Yuditskii study on the probability of rupture of the rocky bed of the river under the action of a jet from a spillway in ski jump, as well as on the estimate of the depth of the erosion pit in the region of impact of the jet. The translation was prepared in 1983 by Pinto de Campos, J.A. at LNEC. It was number 442 and was titled: "Hydrodynamic Action of a Nappe Discharged on Fragments of the Rock Bed of the River and Conditions of Rupture of This."

Yuditskii's work, consolidated in1963, was based on more than 200 tests resulting from years of research on the subject in a hydraulics laboratory. It was the first work developed for rock beds and is the most comprehensive study on the subject.

The research considered the bed of the river without cohesion, with geometric blocks and arranged in a regular form, representing, according to him, a situation more in agreement with reality. The experimental device was assembled in order to allow variation of the parameters: total head, discharge, dimensions of cubic blocks, relative position of the block and relation to the jet impact zone, fracture thickness and depth of excavation.

The jet takeoff angle remained constant and equal to 35°, since it was verified in numerous tests that for $28° \leq \alpha \leq 42°$ there was no appreciable variation of the pressure nor of its fluctuation.

The instantaneous pressures in the block were always measured in steady state. The typical oscillogram is shown in Figure 7.4.

$$B* = P* + A*/2 \qquad (7.5)$$

where:

B^* = instantaneous maximum block pullout pressure (m);
A^* = maximum amplitude of pressure oscillations in the block (m);
P^* = mean value of the vertical head applied to the mass (m).

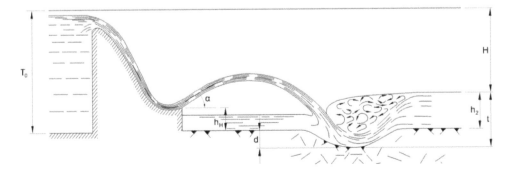

Figure 7.3 Ski jump spillway – schematic sketch.
Source: (Yuditskii, 1963. LNEC, Translation 442, 1983).

Figure 7.3A A photo of the Tucuruí spillway.

T_o = total height of water upstream of the dam;

α = angle of the jet takeoff;

h_H = height of the board lip;

d = depth of the excavation (m);

h_2 = downstream height of water (m);

t = thickness of the water height in the jet impact zone = $d + h_2$ (m);

H = $T_o - h_2$ = gross head (m); Note: in the original paper H = Zo, which appears in Figures 7.6 through 7.8, shows $Z_o = H$;

h_c = critical flow height at the jet takeoff (m);

S = block area (m^2);

a = edge of the cube or height of the block (m);

C = block height (m);

lo = dimension of the block in plan (m); for non-square blocks the dimension in the direction of the discharged liquid vein is considered;

G = block weight (N); and,G^* = submerged block weight = G/S (m).

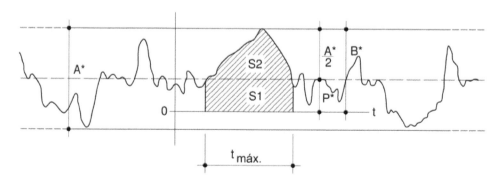

Figure 7.4 Oscillogram

Source: (Yuditskii, 1963. LNEC, Translation 442, 1983).As shown by Figure 7.4 there is the relationship between A^* and B^*:

After the initial test phase, all other tests were performed for the relative positions between the block and jet shown in the following illustration, with fracture thickness equal to 2.0 mm, for which pressure and float values were maximum.

a)   b)

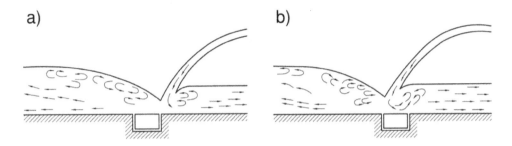

Figure 7.5 Positions relative to block × jet
Source: (Yuditskii, 1963. LNEC, Translation 442, 1983).

During the tests the mentioned parameters varied in the following intervals:

$$6,8 \leq H/h_c \leq 12,9; 0,5 \leq l_o/h_c \leq 3,4; 1,0 \leq h_2/h*_2 \leq 1,5; 0,0 \leq d/h*_2 \leq 2,0;$$

where:

$h*_2 =$ minimum downstream flow height that did not drown the rebound in the jet impact zone

The Froude (*Fr*) and Reynolds (*Re*) numbers relating to the jet launch section varied within the ranges: $10 \leq Fr \leq 150; 2 \times 10^4 \leq Re \leq 10^5$.

Based on values measured at least three times of the maximum amplitude *A* of the pressure oscillations in the block and the instantaneous maximum block pullout pressure *B*, the curves shown in Figures 7.6 and 7.7 relating the dimensionless ratios were plotted:

$$\frac{A^*}{H} = f\left(\frac{t}{h_{cr}}, \frac{lo}{h_{cr}}\right) \tag{7.6}$$

and

$$\frac{B^*}{H} = f\left(\frac{t}{h_{cr}}, \frac{lo}{h_{cr}}\right) \tag{7.7}$$

These approximate curves show that the value of *A** can reach 60 to 70% of the difference of *H* levels and that of *B** can reach 0,5 *H*. The maximum values of *A** and *B** correspond to the situation of the minimum height h_2 downstream.

The tests with non-square plant blocks showed that the values of the maximum amplitude of oscillation of the pressure and of the instantaneous maximum pressure depend only on the transverse dimension of this one. Therefore, these results can also be used for these blocks, simply taking the dimension of the block in the direction parallel to the discharged liquid vein (l_o).

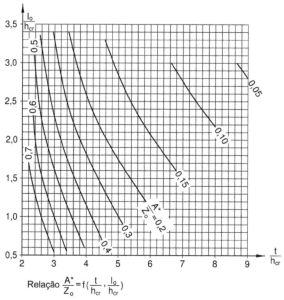

Relação $\dfrac{A^*}{Z_o} = f\left(\dfrac{t}{h_{cr}}, \dfrac{l_o}{h_{cr}}\right)$

A* - Amplitude máxima de oscilação da carga no bloco
(em m de coluna de água)

Figure 7.6 Relation $A^*/Z_o = f\,(t/h_{cr}, l_o/h_{cr})$. In this book, $Z_o = H$ (= A*-maximum load oscillation amplitude in the block (m)).

Source: (Yuditskii, 1963. LNEC, Translation 442, 1983).

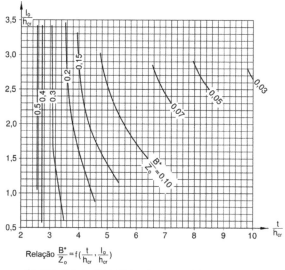

Relação $\dfrac{B^*}{Z_o} = f\left(\dfrac{t}{h_{cr}}, \dfrac{l_o}{h_{cr}}\right)$

B* - Pressão instantânea máxima de arrancamento
aplicada ao bloco (em m de coluna de água)

Figure 7.7 Relation $B^*/Z_o = f\,(t/h_{cr}, l_o/h_{cr})$. In this book, $Z_o = H$ (=B*-maximum instantaneous pullout pressure applied to the block).

Source: (Yuditskii, 1963. LNEC, Translation 442, 1983).

From these data we can construct the representative curves of the relationship between the values of pressure oscillation and the maximum pressure, which are presented in Figure 7.8.

$$\frac{B^*}{A^*} = f\left(\frac{t}{h_{cr}}, \frac{lo}{h_{cr}}\right) \tag{7.8}$$

It can be seen that, for the intervals considered, l_o/h_{cr} and t/h_{cr}, this ratio varies between 0,5 and 1,0, which corresponds to a maximum plucking pressure, varying between 0 (zero) and half of the maximum amplitude of load oscillation, which can be demonstrated by replacing the values quoted in equation 7.8.

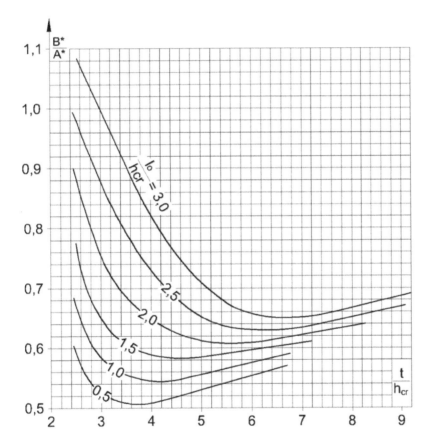

Figure 7.8 Relation B*/A* = f(t/h_cr, l_o/h_cr).
Source: (Yuditskii, 1963. LNEC, Translation 442, 1983).

Conditions for pulling off the block from the rocky bed of the river

For the block to be pulled off, two conditions are necessary: $B *> G *$ and $P *> G *$.

For convenience the weight of the block immersed in water column height: $G_1^* = G / {\gamma_s}$, S being the area of the block.

Yuditskii (1963) concluded, after a series of tests, that the condition of pullout of the block can be expressed by the Equation 7.9.

$$B^*_{lim} = \frac{1,05 \; C \; (\gamma_r - 1)}{1,1 + 0,5 \frac{l_o}{h_{cr}} - \left(0,42 + 0,14 \frac{l_o}{h_{cr}}\right) \frac{A^*}{B^*}} \tag{7.9}$$

With the help of the curves, the value of t/h_{cr} is calculated, from which t and the excavation depth $d = t - h_2$, which is the limit value, are calculated. Adding to this depth, the height of the block (C) has the maximum depth of the excavation: $d_{max} = d + C$.

Finishing this section, it should be noted that the cited LNEC reference presents an exercise of a real case.

7.4   Comprehensive scour model

Important reference should be made to the Comprehensive Scour Model – CSM (Bollaert, 2002, 2004a, 2004b, 2006, 2010, 2012), quoted earlier in Chapter 4, which was presented at the 30th International Conference on Scour and Erosions in Sacramento, California, USSD – United States Society on Dams, in 2010.

Bollaert has several papers on the subject related in the bibliography. In one of the more recent works of 2012, the model was used in the study of the Kariba plunge pool (Chapter 5). As we know Kariba is an arched concrete dam that has a spillway composed of six orifices, with maximum discharge capacity of 8.400 m³/s.

The erosion process evolved over 20 years, as shown in Figures 5.16, 5.17 and 5.19. The scour pit reached 80 m depth in 1981. This erosion was stabilized and presented the same shape in 2001, according to Noret (2012), due to the few spillings in the period and the operation of the nonadjacent gates.

But there was a consensus that erosion could evolve in the event of extreme floods, with the operation of the adjacent gates. The Zambezi River Authority then contracted EPFL's Hydraulic Construction Laboratory (LCH/EPFL) and Aquavision Engineering of Switzerland to assess the future evolution of erosion and to study a new form of the dam that would ensure dam safety (Noret, 2012).

The studies were performed using the mathematical model cited and a physical model fixed at 1:65 scale. The pressures measured in the physical model supplied the mathematical model. The hydrodynamics pressures were compared with the resistance of the rock mass, a good quality gneiss on the left abutment and a quartzite weathered on the right abutment.

The analysis of the results allowed to conclude that the present form of erosion was unstable for the condition of adjacent sluice operations, including a regressive erosion tendency towards the dam foot, as previously discussed in Chapter 5 (Figures 5.16 to 5.19).

Magela (2015a) presented in the 30th SNGB of the Brazilian Committee a summary of the data presented by these authors of interest to all those involved in this type of project.

The Zambezi River Authority, after 50 years of operation of the Kariba plant, was in the process of contracting the work of remodeling the existing pit, as shown in

www.hydroworld.com/articles/2014/10/dam-safety-work-sought-on-plunge-pool-of-africa's-kariba-dam, de 22/10/2014.

It should also be noted that Bollaert published, in partnership with Petry, a work on the spillway of Tucuruí HPP, "Application of a Physics Based on Prediction Model to Tucuruí

Dam Spillway," at the 3rd International Conference on Scour and Erosion in Amsterdam (2006).

The work confirmed the results of studies carried out in the reduced three-dimensional model for the non-cohesive mobile bottom plant project, elaborated by Hidroesb (Rio de Janeiro, Brazil), presented in Chapter 10.

This tool is very important for the design of a structure of this size – see Chapter 6 and Figure 6.4 (Moraes, 1983):

* maximum design discharge: 110.000 m^3/s;
* head: 50 m;
* specific discharge at the launching lip of the jet: 228 $m^3/s/m$;
* rocky massif composed of a metasediments with four different resistances.

The CSM model is totally based on the physics of the erosive process of the rock masses:

* removal of blocks due to fluctuations in joint pressures or shear stresses;
* fracture of rock mass, progressive or sudden;
* abrasion of massive and rock blocks, long-term agents.

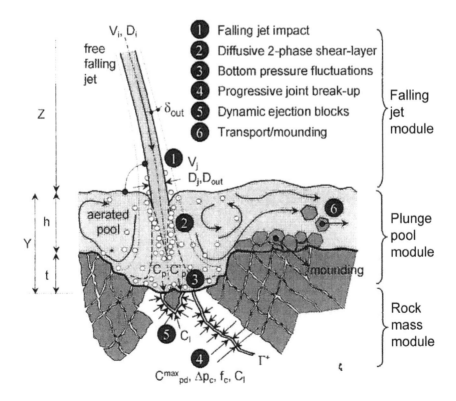

Figure 7.9 Scheme of the physical-mechanical process of scour.
Source: (Bollaert, 2002, 2010)

It is recorded that the model consists of three erosion prediction modules, defined as follows:

*   Dynamic impulse modulus (DI), which expresses the underpressure in the displacement and impulsion of a rock block as a function of the rock density, the size, shape and time of evolution of the instantaneous pressure forces on the block;
*   Mechanical fracturing comprehension module (CFM) which expresses the crushing or growth of subcritical cracking over time as a function of boundary pressure fluctuations, and the type of geomechanical characteristics of the massif;
*   Quasi-uniform impulse modulus (QSI), which expresses the peeling of the thin layers of exposed rock mass as a function of layer thickness, protrusions, block dimensions and shape, and flow velocities near the bottom.

The parameters are presented in previous illustrations, with the exception of Γ'^{+} – amplification factor. The maximum pressure is estimated using Equation 6.11.

The model, available on the internet, composes the author's PhD dissertation presented in EPFL – Ecole Polytechnique Fédérale de Lausanne, in 2002, entitled: "Transient Water Pressures in Joints and Formation of Rock Scour Due to High-Velocity Jet Impact," a subject matter whose consultation is obligatory.

Chapter 8

Cavitation

This chapter presents a summary of the phenomenon of cavitation, covering the conceptualization and characteristic parameters, cavitation caused by irregularities, the protective measures and the specifications for surface protection and several cases of accidents in spillways and tunnels.

8.1 Conceptualization and characteristic parameters

Cavitation is a dynamic phenomenon consisting of the formation and subsequent collapse of cavities (from which the name of the phenomenon originates), or vapor bubbles, in a flowing liquid (Abecasis, 1961; Ball, 1963; Falvey, 1990).

The formation of bubbles occurs in regions where, by any circumstance, the flow accelerates suddenly and the local pressure drops and persists around or below the water vapor pressure (-6.0 m). It is generally verified in cases of separation of the flow.

The collapse occurs because the bubbles are instantly entrained by the flow to zones immediately downstream where the pressure is greater than the water vapor pressure. The sudden collapse of the bubbles gives rise to the creation of very high localized pressures, resulting in fluctuations of pressures, vibrations and noises.

When this phenomenon occurs repeatedly in the vicinity of the solid contour that defines the boundary of the flow, the structure is subjected to shock actions of great intensity.

When the forces resulting from the impact exceed the forces of internal cohesion of the material of the solid surface, its rupture is verified, to which the name of erosion by cavitation is given. This erosion may also be due to a prolonged action of cavitation flow, causing fatigue demands resulting from the repeated action of the phenomenon.

In the case of concrete walls or floors, the destructive action is mainly felt on the less resistant constituent, that is, the binder (binder-cement). Erosion around the aggregate particles increases the roughness of the wall, and conditions for cavitation may become more critical. The particles eventually unleash, and the erosive phenomenon tends to progress downstream. Finding favorable situations, it can reach very significant proportions and cause complete destruction of the lining.

Experience has shown that the cavitation problems in the outlets or spillways, considering the concrete surfaces defining the frontier with suitable lining, take place for speeds greater than 30–35 m/s, depending on this pressure value in the vicinity of the area where the cavitation takes place.

Problems arising at lower speeds are due to imperfect surface lining or local turbulence caused by slots in gates, columns, spreading blocks, damping and falling blocks, etc., or by reducing the pressure caused by the layout of the works.

The onset and intensity of cavitation depend on the dynamic structure of the flow characterized by the velocity distribution, the characteristics of the boundary layer and the mean pressures and pressure fluctuations.

If the forces of viscosity and compressibility can be neglected, the pressure varies with the square of the velocity, and the characteristic parameter of the cavitation (σ) has the expression:

$$\sigma = \frac{p - p_v}{V^2/2g} \tag{8.1}$$

where:

p = absolute pressure at a point in the system, outside the cavitation zone (m);

P_v = vapor pressure inside the cavitation bubbles, which is usually taken equal to the water vapor pressure at the temperature at which it is present (m);

V = flow velocity at a flow reference point (m/s);

g = acceleration of gravity (m/s^2).

The critical parameter, σ_{CR} or σ_I, which characterizes the start of cavitation in various hydraulic structures, for different operating situations, is obtained experimentally in reduced models. The practical results for design effect are presented in the next section for abrupt and gradual irregularities. It is necessary, however, to take into account the factors that can affect the pressure field and influence the transposition of the results into prototypes. For details, it is recommended to consult Quintela and Ramos (1980).

8.2   Cavitation caused by irregularities

Ball (1963) carried out an extensive program of experimental investigations to define incipient conditions of cavitation (σ_{CR}) for abrupt and gradual gradients along the solid contour of hydraulic surfaces. He classified the main irregularities observed on the solid surfaces, indicating the corresponding regions affected by cavitation erosion, which are shown in Figure 8.1.

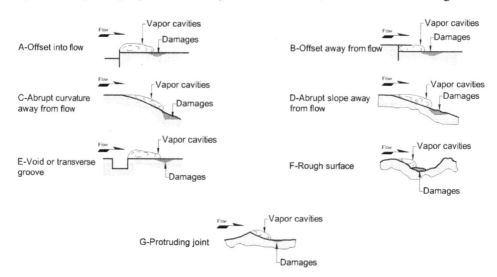

Figure 8.1 Types of irregularities and damage areas by cavitation.
Source: (Ball, 1963).

The recommended results to guide the finishing projects of these surfaces are presented in Figure 8.2, which was extracted from Publication 35, CEHPAR, "Cavitation and Aeration in High Velocity Flows" (Pinto, 1979).

In this illustration, the results (σ_{CR}) are observed for an abrupt shoulder and a recess, also abrupt, positioned transversely to the direction of the flow. One can also observe the result (σ_{CR}) for a relaxed surface. The favorable effect of the easing of these irregularities is evident. A step of 0,70 cm of lowered height, in the ratio 20 (H): 1 (V), has a value of σ_{CR} 10 times smaller (2,05 to 0,2).

It should be noted that, in these studies, no mention is made of the dynamic structure of the flow, characterized by the velocity distribution, the boundary layer characteristics, and the fields of medium pressure and pressure fluctuations.

It is known that the development of the boundary layer is the main cause of the better performance of the spillways when compared to the bottom discharges. In these structures, the steep flow curvature downstream of the gate eliminates the boundary layer, and any millimetric irregularity is subjected to very high velocities.

In spillways, the increase of flow velocity is done concomitantly with the development of the boundary layer. Small irregularities of the order of centimeter can be tolerated. For a more detailed analysis of these aspects, it is recommended to refer to the references cited.

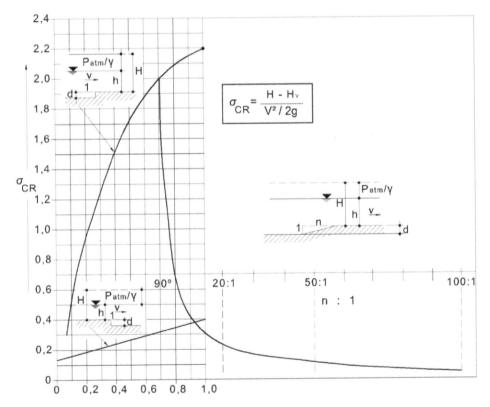

Figure 8.2 Incipient cavitation σ_{CR} values for offset "d" (cm); abrupt and gradual on the contour surface.

Source: (Ball, 1963; Pinto, 1979).

On the same subject, it is necessary to present the graph of the Incipient Cavitation Index of the NUST (2003), Figure 8.3.

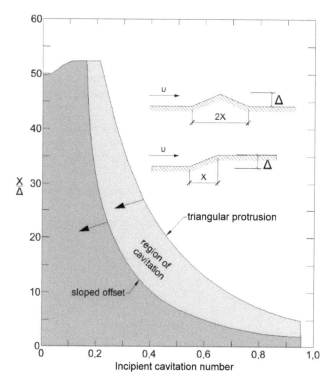

Figure 8.3 Values of σ_{CR} for incipient cavitation for small protrusions on the contour surface. *Source:* (NUST, 2003).

8.3   Protective measures – surface finish specifications

The most frequent measure for protection of the concrete surfaces of the discharge organs submitted to free surface flows with high speeds against cavitation erosion caused by irregularities has been the adoption of more stringent specifications for the finishing of these surfaces.

Abrupt irregularities on these surfaces are not tolerated. The specifications make it necessary to reduce them to inclined surfaces by bevel grinding, depending on the speed of the flow.

In addition, for the most vulnerable sections of surfaces, it has been recommended to use more resistant materials – special concrete replacing the steel armor, until recently considered by some specialists as the ideal but costly and impractical on very large surfaces.

In the stretches of the discharge devices with pressure flow, erosion by cavitation can in many cases be avoided by securing a suitable cross-section to the conduit. By increasing the section of the conduit, the speed is reduced, and consequently the pressure is increased. These aspects combine in the favorable sense of the non-appearance of cavitation.

The specifications for finishing the surfaces of the USBR establish that:

- recesses or depressions are limited to 3,2 mm and 6,4 mm, depending on whether the surface is formed or not and whether the irregularity is transverse or in the direction of flow; when the irregularity is located downstream of a gate, it must be completely removed;
- unacceptable irregularities are removed or reduced to an appropriate size by bevel grinding. The slopes must be less than 20 (horizontal):1 (vertical), 50 (H):1 (V) and 100 (H):1 (V) for speeds of 12 to 27 m/s, 27 to 36 m/s and greater than 36 m/s, respectively. Slopes 50 and 100 (H):1 (V) are impractical and, in such cases, it is necessary to use aeration.

The finishing conditions imposed by any of the above values are extremely restrictive and therefore of difficult and costly execution, since the treatment of the irregularities is done by grinding the surfaces.

By analyzing Ball's curves in detail, it is verified that for velocities of the order of 35 m/s and absolute pressure of the order of 5 m of water column, surface irregularities should be ground in the ratio 100 (H):1 (V).

Several experts consider that it is materially impossible to comply with this restriction. It can be seen that, for the mildest slopes, between 50 (H): 1 (V) and 100 (H): 1 (V), no significant reduction of σ_{CR} is obtained, and therefore, such a decision would not be technically and economically justified.

In these conditions, and considering the executive difficulties of special concretes in large areas, the most logical alternative is the aeration of the liquid vein.

For an analysis of this aspect, it is recommended to consult references already cited, including Neidert (1980a).

For the detailed analysis of cavitation erosion in slots, dispersion blocks and stilling basin block, and for details on the use of more resistant materials (shields, concrete with steel fibers, epoxy mortars, use of resins and impregnation of polymers), one should consult Quintela and Ramos (1980), which includes the Russian experience of several authors, who admit larger heights for irregularities, as shown in Table 8.1. In Table 8.2, the values of σ_{CR} for transverse and longitudinal irregularities as a function of slope are presented.

Falvey (1990), in the USBR Engineering Monograph 42, Cavitation in Chutes and Spillways, presented project recommendations to which are transcribed in Table 8.3.

This publication presents important considerations for the geometric designs of chute spillways and tunnels. As the flow dips into the downward branch of the profile, the velocity increases and the depth of water is reduced. These effects lead to low cavitation rates and increase the possibility of damage.

If the structure does not have a vertical curve, which can be used to control the cavitation index, then the flow depth must be controlled by reducing the chute width or the tunnel diameter. This section change must be made with care in such a way that the tunnel is not flooded and, in case of a chute, the side walls are not overtopped.

According to Falvey, few spillways were designed considering primarily the characteristics of cavitation. One exception was the Aldeadavila HPP Project (Figures 8.4 to 8.6), 1.200 MW, eight machines, built between 1956–1962 on the Spanish section of the Douro river. The site of the development is a "V" valley, consisting of granite, with more than 400 m of depth.

Table 8.1 Inclination of chamfers to prevent erosion by cavitation.

(Oskolov and Semenkov, 1979).

Head (m)	Height of Irregularity (mm)	Inclination of chamfers/Irregularity Transversal		Longitudinal
				-
40	5	-	-	-
to	5 to 10	1:4	1:8	1:2
50	10 to 20	1:8	1:10	1:3
	20 to 40	1:12	1:14	1:3
60	2,5	-	-	-
to	2,5 to 5	1:7	1:11	1:2
70	5 to 10	1:14	1:18	1:3
	10 to 20	1:16	1:20	1:3
	20 to 40	1:20	1:24	1:3
80 to 100	10 to 20	1:32	1:38	1:4
	20 to 40	1:36	1:42	1:4

Table 8.2 Parameters of erosion by cavitation – σ_{CR}

(Oskolov and Semenkov, 1979).

Inclination d/a	Inclination of chamfers/Irregularity Transversal		Longitudinal
			-
1:2	0,73	0,77	0,40
1:3	0,70	0,74	0,30
1:4	0,67	0,71	0,23
1:6	0,61	0,66	-
1:10	0,52	0,57	-
1:14	0,45	0,49	-
1:18	0,39	0,43	-
1:22	0,34	0,38	-
1:26	0,31	0,34	-
1:30	0,28	0,31	-
1:34	0,25	0,28	-
1:38	0,23	0,26	-
1:40	-	0,24	-

The project discharge is 12.500 m³/s; the head is 124,50 m; the dam is 250 m long and 139,50 m high; width of the river channel is 60 m; the level of water downstream is negligible because the profile of the river is very steep. The alternative of arch dam prevailed, with a spillway over the dam with eight spans 14 m wide × 7,80 m high, with a capacity of

Table 8.3 Tolerances for finishing of concrete surfaces submitted to the flow (Falvey, 1990)

Tolerance, T	Offset (mm)	Inclination
T1	25	1:4
T2	12	1:8
T3	6	1:16

Table 8.4 Specifications of tolerances of concrete surfaces submitted to the flow (Falvey, 1990).

Cavitation index	Tolerance without aeration	Tolerance with aeration
> 0,60	T1	T1
0,40 to 0,60	T2	T1
0,20 to 0, 40	T3	T1
0,10 to 0,20	Review Project	T2
< 0,10	Review Project	Review Project

Figure 8.4 Aldeadavila HPP. View of the dam. Douro river, Spain.

10.000 m³/s, complemented by a tunnel spillway with a capacity of 2.800 m³/s. Hydraulic studies in a reduced model lasted three years.

During construction, in January 1962, the dam was overtopped without serious damage. In the tunnel there was great cavitation damage of several concrete ring joints adjacent to a failure; 50 m of lining was destroyed, which was being repaired (in 1967). Two fixed-cone dispersion valves were installed, which operated for six months at maximum discharge and broke.

The rupture was corrected, the aeration design was increased and the operation was satisfactory (Falvey, 1990). For more details, it is recommended to consult Galindez (1967).

Figure 8.5 Aldeadavila surface spillway. Capacity 10.000 m³/s.

Figure 8.6 Aldeadavila tunnel spillway. Capacity 2.800 m³/s.

The project script presented by Falvey (1990: 77), in summary, prescribes:

1   Based on the contour conditions, decide on the complete preliminary draft of the spillway and study the cavitation characteristics of the structure.
2   Determine the surface finish tolerances as a function of the cavitation index. Table 8.4 indicates whether aeration is desirable.

3　If there are cavitation indices less than 0.20, the effects of a change in spillway geometry in the cavitation index should be investigated.
4　If the index cannot be improved by the changes of geometry, it is recommended to use aerators.
5　For values of very small cavitation indices ($\sigma < 0,10$), changes in the design concept should be considered.

It is worth recording the project of the Tucuruí HPP spillway. Intensive research was done on the possibilities of cavitation occurring in this spillway where, at some points, maximum velocities of the order of 30 m/s could occur. This research was consolidated by the Eletronorte Report (TUC-10–7365-RE of June 1982).

The technical specifications for finishing the concrete surface allowed gradual irregularities of a maximum of 1:50, with $d = 0,6$ m. From Figure 8.2 we have a value of the critical cavitation index σ_{CR} of 0.10 for irregularities with this inclination and $d = 0.7$ cm. For $d = 0,6$ cm, this value is lower.

The values surveyed for the Tucuruí spillway over the whole extension of the slope profile, for the different operating conditions (flow with and without floodgate control), were always higher than $\sigma_{CR} = 0,40$. In the region of the end of the shell, the minimum values reached 0,15, still with safety in relation to the σ_{CR}.

The spillway has been operating since October 1984 without problems, as reported by Magela et al. (2015): "Thirty Years of Operation of the Tucuruí HPP Spillway." The areas damaged by cavitation downstream of the stoplog groove are shown in Figures 8.7 to 8.9.

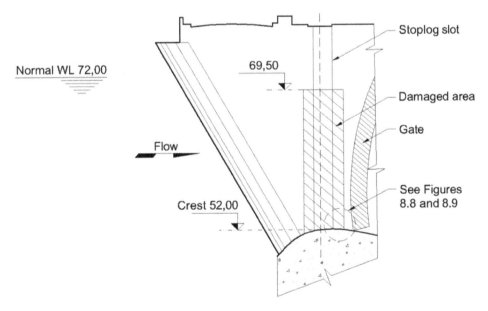

Figure 8.7 Spillway of Tucuruí HPP. Area damaged by cavitation.
Source: (CBDB/LBS, 2002).

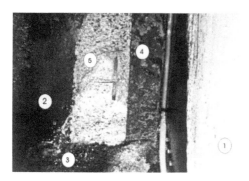

Figure 8.8 Damaged area downstream of the stoplog slot.

Figure 8.9 Repairs performed.
Source: (CBDB/LBS, 2002).

8.4   Cavitation cases

The following are cases of some notable spillways where cavitation problems occurred. It is worth mentioning the work of Oliveira *et al.* (1985a), in which they showed the cases of erosion of the concrete surface of the spillways of Ilha Solteira (34.300 m^3/s) and Jurumirim (2.950 m^3/s), as well as describing the repairs performed.

8.4.1   *Spillway of Porto Colombia*

Porto Colombia HPP (320 MW), owned by Furnas, is located in the Grande river, in the border of the states of Minas Gerais and São Paulo, Brazil (Figures 8.10 to 8.15). Its operation began in 1973. The spillway, with six radial gates of 15 × 15,4 m, has a maximum capacity of 16.000 m^3/s. The energy dissipation is made by a hydraulic jump in a stilling basin.

Figure 8.10 Stilling basin of the Porto Colombia HPP spillway.
Source: (CBDB, 2002).

Figure 8.11 Cavitation downstream of the chute blocks and in the slab of the dissipation basin. Holes 2m × 1,7 m × 0,7 m deep.

Figure 8.12 Cavitation in the kick blocks, as well as in the stilling basin slab.
Source: (CBDB, 2002).

Figure 8.13 Cavitation in the end sill blocks.
Source: (CBDB, 2002).

Underwater inspections were carried out in 1983 and 1990. Problems were detected in the chute blocks and in the end sill. For details, see the publication *Large Brazilian Spill-ways* (CBDB, 2002). There are several other cases of this type of problem reported by Berryhill (1957, 1963).

The solution adopted by Furnas to solve the problem was to eliminate the chute blocks and to make a continuous end sill, as shown in the following photos.

Seven pressure cells were installed in the prototype in 1996 in order to measure the hydraulic efficiency of the modifications, as well as to compare the results with those of the model. The results are reported in Furnas report of December 1996 that the author did not have access to.

Figure 8.14 Stilling basin without the chute blocks.
Source: (CBDB, 2002).

Figure 8.15 Stilling basin with continuous end sill.
Source: (CBDB, 2002).

8.4.2   *Spillway of Shahid Abbaspour*

This plant (2.000 MW), the former Karun 1, is located on the Karun river, Iran (Figures 8.16 to 8.18). The chute spillway on the left abutment has a flow capacity of 16.000 m³/s (World Water, 1979). In 1978, cavitation occurred in a stretch of the spillway submitted to a head of 100 m, consequently, with a speed of the order of 43 m/s. A part of the chute slab with

Figure 8.16 Shahid Abbaspour HPP.
H = 200 m (1976).

Figure 8.17 Shahid Abbaspour HPP. Spillway operating.
Source: (Pinto, 1979).

Figure 8.18 Shahid Abbaspour HPP. Cavitation in spillway chutes.
Source: (Pinto, 1979).

irregularities on the surface was eroded, as shown in Figure 8.18. Details are presented by Pinto (1979).

8.4.3 *Spillway of Guri*

The Guri HPP (10.235 MW) is located in Caroni river, Venezuela (Figure 8.19). The dam is 162 m high. The spillway, with nine gates of 20 m × 15 m, has a flow capacity of 27.000 m^3/s and was operated nine months a year from 10 to 14.000 m^3/s discharges (Chavarri *et al.*, 1979). At the jet takeoff lip, under a head of 120 m, the velocity was of the order of 47 m/s and negative pressures of up to −5,0 m occurred. Under these extreme operating conditions cavitation erosion occurred, reaching depths of 4 to 7 m (Figure 8.20). The repair work included changing the bucket profile of the terminal sill. The studies were done in the laboratory of Saint Anthony Falls of the University of Minnesota (Minneapolis, USA).

festivalmada.com.b

informepastam.com

Figure 8.19 Guri HPP – 1978.

Figure 8.20 Guri HPP. Jet takeoff lip cavitation detail.
Source: (Chavarri, 1979).

8.4.4. *Spillway of Dworshak*

This HPP (400 MW) is located on the North Fork river, Clearwater, Idaho, USA (Figures 8.21 and 8.22). The dam is 212 m high and 1.002 m long. The spillway, in hydraulic jump, has two gates and a flow capacity of 4.000 m^3/s. The dam additionally has three half-bottom dischargers, with dimensions 3,7 m × 5,2 m, with a capacity of 1.180 m^3/s under a head of 81 m. The maximum difference between the maximum reservoir level and the basin level is

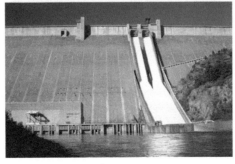

Figure 8.21 Dworshak dam. USCE, Walla Walla District.
Source: (Kollgaard and Chadwick, 1988).

Figure 8.22 Dworshak dam. Erosion in the stilling basin. US Corps of Engineers.
Source: (Regan, 1979; Pinto, 1979; Neidert, 1980a).

208 m. Under this head, the velocities at the entrance of the basin are on the order of 60 m/s. Both the basin and the bottom dischargers had serious cavitation problems.

8.4.5 *Spillway of Libby*

This HPP (600 MW) is located on the Kootenai river, Montana, USA (Figures 8.23 and 8.24). The dam is 136 m high and 684 m long. The spillway in hydraulic jump has two

Figure 8.23 Libby dam. US Corps of Engineers.

Figure 8.24 Abrasion Libby dam stilling basin.
Source: (Neidert, 1980a; Regan, 1979).

gates and a flow capacity of 4.100 m³/s. It has, additionally, three half-bottom dischargers, with dimensions of 3,7 m × 5.2 m, to 1,698 m³/s under 81 m of load. The maximum head between the maximum level of the reservoir and the level of the basin is 118 m. Under such conditions, the maximum flow velocity at the entrance of the basin is of the order of 48 m/s. As in Dworshak, both the basin and the bottom dischargers had serious cavitation problems.

8.4.6 *Spillway of Keban*

This plant (1.240 MW) is located on the Euphrates river in the Turkey (Figures 8.25 and 8.26). The construction was completed in 1973. The chute spillway with a ski jump has 6

Figure 8.25 Keban HPP. *H* = 158 m. Note: downstream has the dams of Karakaya and Atatürk.
Source: (Wikipedia).

Figure 8.25 (Continued)

Figure 8.26 Keban HPP – Cavitation in spillway chute.
Source: (Aksoy 1979; Demiröz and Ivazoglu, 1991).

sluices of 24×16 m and a flow capacity of 1.700 m³/s. The head between the crest and the bucket is 106,6 meters. Therefore, the flow velocity is on the order of 44 m/s. The operation of the spillway began in 1979, after filling the reservoir. Soon after, the cavitation occurred in the chute due to irregularities in the concrete surface (offsets) in the region the joints between the concrete slabs (Aksoy and Ethembabaoglu, 1979)

8.4.7 *Service spillway of Tarbela*

The Tarbela HPP is situated in the Indus river, Pakistan (Figure 8.27 A). In the left chute of the spillway bucket, erosion by cavitation occurred (Figure 8.27 B). According to Lowe

Figure 8.27A Tarbela HPP. Service spillway.
Source: (Dost Pakistan).

Figure 8.27B Cavitation in the left chute, bucket 4, of the service spillway of Tarbela.
Source: (Lowe, 1979).

(1979), the spillways and the dam tunnel were used many months each year at maximum capacity and flow velocities in the plunge pools reached 44 m/s (Figures 5.11 and 5.11A). Reference should also be made to the paper of Binger (1972).

8.4.8 *Service and emergency spillways of Oroville*

The Oroville HPP (Figures 8.28A to 8.28G), on the Feather river, was inaugurated in 1967 and is owned by the California Department of Water Resources – DWR. It is located 120 km north of Sacramento and 8 km east of the city of Oroville. Its objectives are water supply, power generation, flood control and recreation. The powerhouse on the left bank is under-ground, 675 MW, with three Francis turbines (487 m^3/s) and three modified turbines (158 m^3/s). The maximum discharge recorded at the site was 7,530 m^3/s (Kollgaard and Chadwick, 1988).

Figure 8.28A Oroville dam.

Figure 8.28B Right bank view.

Figure 8.28C Damage to the service spillway.

Figure 8.28D The broken slab (90 m × 150 m × 15 m).

Figure 8.28E Emergency spillway running. Erosion in the hydraulic left at the foot of the sill with the possibility of regressing and collapsing the spillway.

The earth dam is 235 m high, crest at elevation 281 m and water level absolute maximum at elevation 274,30m. The free board of 6.7 m is intended for flood control. The dam is founded on a rocky metavolcanic formation. The massif is predominantly amphibolite,

Figure 8.28F Emergency spillway. Repair works with rock blocks.

with abundant veins of calcite, quartz, asbestos and pyrite. It is hard, dense, grayish-green to black, fine to thick granulation, moderately to strongly fractured, and cut transversely by shear zones and schistosity with sharp dipping. Three prevailing sets of joints impart a fracture to the rock, but individual joints are relatively narrow. Two larger shear zones exist beneath the dam in the middle of the height at the abutment. Both dive heavily and attack the axis of the normal dam.

The plant has two spillways on the right abutment with a total flow capacity of 17.700 m³/s:

- service gated spillway, for flood control, with eight gates of 5.2 × 10m and capacity for 2.530 m³/s (C = 2 and H = 14 m), with a rapid of 950 m until the downstream 167,60 m elevation; therefore, with 106,70 m of unevenness, it can be inferred that, at the end of the rapid, the flow velocity will be of the order of 45 m/s, that is, very high;
- emergency spillway, a free overfall with 527,3 m extension, including parking on the right jamb (thick sill) and capacity of 15.170 m³/s (H = 6,4 m and C = 1,8), without slab protection of the downstream massif; therefore, it was decided to live with the erosion in case of operation of this structure.

On February 7, 2017, an incident occurred with the service spillway when it was spilling 1.400 m³/s. The spillway was closed for repairs and the emergency spillway came on stream for the first time in 50 years. Soon after the operation, on February 11, erosion occurred downstream of it. It is recorded that in 2005 the construction of a concrete track was evaluated downstream of the emergency spillway on the slope of the river, which, for cost reasons, was not implemented.

The record in this book of this incident is merely informative. It was prepared based on official information from the DWR website and project information from the USCOLD Development for Dam Engineering in the United States prepared for the 16th ICOLD in San Francisco, edited by Kollgaard and Chadwick (1988).

After many years of drought, in February 2017 there was a flood of 2.800 m³/s, and both spillways worked. Extensive damage occurred on the service spillway slab and erosion occurred downstream of the emergency spillway. The sequence of the incident was as follows:

- occurrence of displacement in the bottom slab of the service spillway, possibly by cavitation; the high velocity flow may have attacked irregularity of some joint or crack in the slab, beginning an erosive process;
- the service spillway was closed for evaluation of its recovery, and the level of the reservoir was rising due to heavy rains in the upstream basin; the DWR engineers concluded that it would be impossible to recover the service spillway chute;
- in anticipation of more precipitation, it was decided to extravasate the flood by the emergency spillway; in a few hours there was erosion in the hydraulic left, at the foot of the sill, with the possibility of breaking this spillway, according to the DWR, and making the situation uncontrolled;
- these facts led to the decision to evacuate about 200.000 inhabitants in the downstream valley and to resume operation of the service spillway, which, although damaged, was necessary to demount the reservoir, with a reduction of the clearance through the free sill;
- restarting of the spillings by the main service chute aggravated the damages where the initial displacement occurred, but still distant from the dam, and with the tendency to progress with greater intensity downstream.

On February 22 the situation was controlled: the reservoir's water level was lowered and the recovery work continued.

The aspects summarized in Figure 8.28 G show that the possibility of the dam collapsing as a result of this incident was remote.

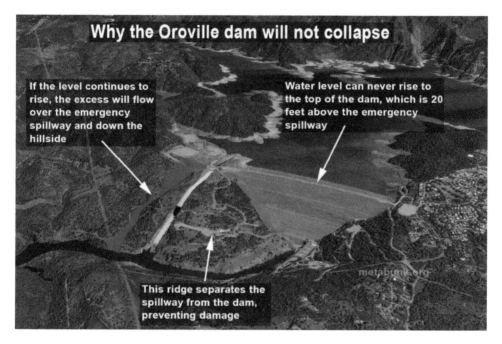

Figure 8.28G Why the Oroville dam will not collapse.
Source: (metabunk.org).

Exploring this case a bit more, in the author's opinion, it can be assumed that the initial damage of the service spillway was caused by cavitation, as occurred at the Shahid Abbaspour dam (formerly Karun 1) on the Karun river (Iran), described in section 8.4.2, and in other incidents previously presented.

As mentioned earlier, the flow velocity at the end of the chute is of the order of 45 m/s, very high. Any irregularity, crack, transverse to the flow, can cause the cavitation to begin. At Guri spillway, a large cavitation accident occurred at the jet's takeoff lip (section 8.4.3).

Finally, it should be mentioned that, in 2013, cracks occurred in the slab of the chute that were repaired.

Previously, in 1975, during a seismic shock in California of 5,7 degrees on the Richter scale, the dam resisted the tremor with little damage, as reported in the event promoted by ASCE in 1976, The Evaluation of Dam Safety, Asilomar Conference Grounds Pacific Grove, California.

8.5   Cavitation cases in culvert (bottom outlets)

On the subject is recommended to consult among others: Engineering Monograph 42 USBR, Cavitation in Chutes and Spillways, by Falvey (1990). This monograph gives to the designer of hydraulic structures both an understanding of cavitation and the design tools necessary to eliminate or reduce the damaging effects of cavitation in chutes and spillways.

The monograph discusses basic concepts, cavitation damage, practical methods of coping with cavitation damage, design recommendations, the influence of geometry and aeration, and the Bureau of Reclamation field experience in the Chapter 7, with six examples: Blue Mesa dam in 1970 (Gunnison river, Colorado), Flaming Gorge dam in 1975 (Green river, Utah), Glen Canyon dam in 1983 (Colorado river, Arizona), Hoover dam in 1942 (Colorado river, Nevada), Kortes dam in 1983 (North Platte river, Wyoming) and Yellowtail dam in 1967 (Bighorn river, Montana).

It is recommended to consult also Pinto (1979), which presents some examples where accidents due to cavitation occurred: Palisades, Hoover and Yellowtail dams (USA), Serre-Ponçon dams and Sarrans (France), San Esteban and Aldeadávila (Spain), and Infiernillo (Mexico).

In the Serre-Ponçon dam (380 MW) outlet number 1 (see Figures 8.29 to 8.32), which had a maximum static pressure of 124 m and a velocity of 45 m/s, there was an important scour of the concrete bottom downstream of the armored section on two occasions: August 1959 and December 1960 (Figures 8.31 and 8.32).

The first time, after a 40-day run, at a discharge of 155 to 415 m^3/s, under a head of 85 m, scour reached a volume of 360 m^3. After performing the repair, the outlet has returned to service. When the discharge reached 300 m^3/s, at the end of a day of operation, scour occurred again. The final repair was done again with concrete in order to obtain a roughness equal to that of the metallic shield. Tests carried out with discharges of up to 600 m^3/s during five hours did not reveal any appreciable erosion.

Figure 8.29 Serre-Ponçon HPP. Durance river, Provence, France.Note on the left abutment two outlets with capacity of 1.700 m^3/s and tunnel spillway for 1.800 m^3/s.

The following is a summary of the case of the Glen Canyon concrete dam (Figures 8.33 to 8.36), 216 m high, founded on sandstone. This dam has a crest with a length of 474 m, a crest with a width of 7,6 m and a base with a width of 91,4 m. It has a bottom outlet on the left abutment, for 425 m^3/s, with four hollow-jet valves.

In addition, there are two tunnel spillways, one at each abutment, 12.5 m in diameter, lined with concrete – 1,8 m thick, with a total capacity of 2,900 m^3/s, controlled by

Figure 8.30 Serre-Ponçon. Tunnel spillway operating.
Source: (Commons Wikipedia).

Figure 8.31 Serre-Ponçon outlet. Scour downstream from the bottom gate.
Source: (Pinto 1979).

radial gates 12,2 m × 16,0 m. The profiles of the tunnels have inclined sections of 55°, vertical curve (elbow), horizontal section of 305 m and in the end a flip bucket – see Figure 8.33 (Kollgaard and Chadwick, 1988).

Both tunnels operated frequently in 1980. The 1981 inspection identified some cavitation damage, larger in the left tunnel due to longer operating time. Protection–aeration

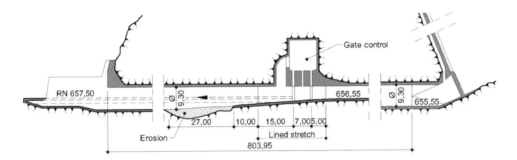

Figure 8.32 Serre-Ponçon outlet.

Source: (Pinto 1979).

measures planned to be carried out in 1984 were studied. But the flood of June 1983 caused extensive damage in both tunnels. Figure 8.35 shows the "big hole" of the left tunnel.

The repairs were performed in 1984, when the aeration slots were implanted, as shown in the Figure 8.36, upstream of the beginning of the vertical curve where the damage began.

In very eroded areas the damaged concrete was removed, the hardware was reinforced and the surface was concreted. Some areas at the end of the elbow have not been repaired. In this case, the experience acquired in the Yellowtail dam repair criteria was used.

After the repairs several tests were performed and no further damage was observed. For details see Falvey (1990).

In the Hoover dam (Chapter 2) there was a similar case of cavitation of the 15,2 m diameter spillway tunnel of the Hoover dam, Arizona side. The hole is 35 meters long, 9 meters wide and 14 meters deep.

Figure 8.33 Glen Canyon (1.021 MW). Colorado river. Bottom outlets and spillway operating.

Source: (Kollgaard and Chadwick, 1998).

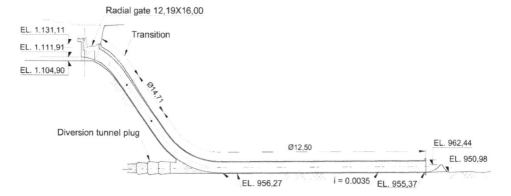

Figure 8.34 Longitudinal profile of the Glen Canyon tunnels
Source: (Kollgaard and Chadwick, 1988).

Figure 8.35 Glen Canyon. (Kramer, 2004). "Large hole" in the elbow region with 11 m depth caused by cavitation.

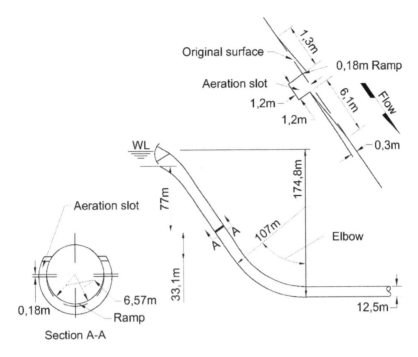

1,3m

Original surface

0,18m Ramp

Aeration slot

Flow

1,2m

6,1m

1,2m

0,3m

WL

174,8m

77m

Aeration slot

A A

107m

Elbow

33,1m

6,57m

0,18m

Ramp

12,5m

Section A-A

Figure 8.36 Glen Canyon tunnel spillway: Aeration slots.
Source: (Falvey, 1990).

A

Figures 8.36a and b Hoover tunnel spillway.
Source: (Falvey, 1990).

Figures 8.36a and b (Continued)

8.6 Aeration

According to Pinto (1979, 1983, 1989), a concrete surface is not free from cavitation damage when flow velocities reach or exceed 40 m/s, as well as for high specific discharges, even when performed to specifications rigid finishing. For high specific discharges, common in current projects, the entrainment of air from the surface of the water due to the development of the boundary layer does not always reach the region of the channel bottom. In the absence of the protection provided by the water-air emulsion any surface irregularity capable of reducing the pressure locally to the level of vaporization of the water becomes important. The greater the flow velocities, the greater the possibilities of cavitation. For velocity equal to or greater than 40 m/s, the pressure field is particularly sensitive, as shown in Figure 8.37. It is shown that a speed increase of 5 to 10% is sufficient to cause pressure reductions of 10 m of water column.

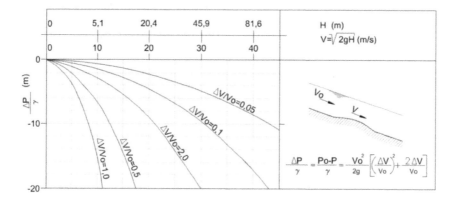

Figure 8.37 Local pressure reduction as a function of speed variation.
Source: (Pinto, 1983).

Experience shows that it is very difficult, or virtually impossible, to build concrete surfaces with a smooth enough finish to prevent cavitation at these speeds. In these cases, steps or ramps can be used to promote the admission of air (artificial aeration) under the lower nappe of the flow, directly on the contact surface; this appears as a logical alternative to the adoption of very smooth and regular surfaces, whose execution is difficult and expensive and not always successful.

Several dams were built with aeration devices: Bratsk, Ust-Ilim, Nurek and Toktogul-Narim in Russia, Yacambu in Venezuela, Itaipu and Foz do Areia on the Iguaçu river, Paraná (Brazil), and Emborcação on the Paranaíba river, Minas Gerais (Brazil).

Designing an aeration system requires answering three main questions (Pinto, 1983):

1   At what speed should the first aerator be provided?
2   What is the volume of air entrained in the aerator?
3   What is the space between the aerators to maintain a given level of protection?

The answer to the first question is related to the concept of aeration as a system of protection against cavitation. The care taken during construction certainly helps reduce the irregularities and, consequently, the risks of cavitation.

The following table presents incipient cavitation index values (σ_i), which indicate the effect of surface quality. Quality improvement is naturally related to the cost increase due to the difficulty of performing surfaces with more stringent tolerance levels.

Table 8.5 Incipient cavitation indices (see Figure 8.2).

1: n (n = slope of ramp)	σ_i ($\sigma_i = (H-H_v)/V^2/2g$)
1: 5	0,62
1:10	0,50
1:20	0,35
1:40	0,25

The basic idea underlying the aeration systems is the reduction of costs resulting from less rigorous specifications for the finishing of concrete surfaces. An adequate evaluation of the incipient cavitation index (σ_i) for the irregularities that are expected in the work and economic considerations should support the decision of where to place the first aerator. Available information on the effects of aeration (Pinto, 1983) indicates that an air concentration of 5 to 10% ($C= [Var/Var+Vw]$) near the surface to be protected practically eliminates the risks of cavitation (Pinto, 1983). However, a suitable design of an aeration system depends on the correct estimation of the amount of air to be drawn by the aerator (second question) and the development of an air concentration close to the surface to be protected (third question). Additional aerators should be provided in sections where the concentration falls below the required minimum level.

Ramps, steps, recesses, slots, etc., inserted in the chute of the spillway are the devices used to cause a natural aeration of the flow next to the concrete surface. The sudden discontinuity in bottom alignment creates an air-water interface in which the high velocity of water entrains air in an intense mixing process. The definition of aerator geometry and its influence on aeration performance is not an easy task. For details on the subject, specifically on the topics air drag mechanism, dimensional analysis, prototype results and conformance model × prototype, it is recommended to consult the works of Pinto (1979, 1983, 1989). The following is a summary, extracted from Pinto (1989), based on measurements in the prototypes of Foz do Areia HPP, Emborcação HPP, Amaluza HPP, Colbun HPP and Tarbela HPP.

8.6.1 Design of an aeration system – summary

The geometry of the aerator is defined in Figure 8.38.

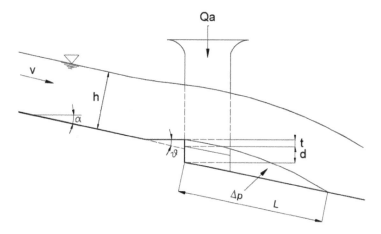

Figure 8.38 Geometry of the aerator

Source: (Pinto, 1983). where: α = angle of inclination of the chute; ϑ = ramp angle relative to the bottom of the chute; t = ramp height (m); B = width of the chute, corresponding to each aerator (m); A = duct area under the nappe (m²); Q = total water flow (m³/s); q = specific discharge of water per unit width of the chute (m³/s/m); Q_a = total air discharge (m³/s); q_a = specific air discharge per unit width of the chute (m³/s/m); h = water height in aerator section (m); V = average flow velocity: $Q = q/h$ (m/s); Fr = Froude number: $Fr = V/(gh)^{0.5}$. The air duct is considered as a short orifice or conduit, defined by its cross-section and consequent loss of charge of the air flow. D = effective duct area per unit of chute width: $D = CA/B$ (m²/m); C = discharge coefficient for the air discharge formula through ducts.

$$q_a B = CA \sqrt{2 \frac{\Delta p}{\rho_a}}$$ (8.2)

where:

Δp = the difference between the atmospheric pressure and the average pressure under the jet, as measured along the vertical face of the step or the ramp (N/m²)

ρ_a = the density of the air (kg/m³).

The performance of the aerator is measured by the ratio of air and water discharges:

$$\beta = \frac{Qa}{Q} = q_a/q$$ (8.3)

In general, as quoted by Pinto (1991) in "Modeling Aerator Devices – Dimensional Considerations," XXII Congress IAHR, Lausanne, Switzerland, the performance of an aerator, insofar as the drag effect is concerned, can be represented by a function:

$$\beta = f(Fr, t/h, D/h) \tag{8.4}$$

where:
- the Froude number, Fr, reflects the conditions of the water flow;
- t/h is a measure of the relative height of the step or ramp;
- D/h characterizes the aerator effect.

When the dimensionless parameters Fr, t/h, D/h are uniquely dependent on the water flow and β is a unique function of the water flow, as shown in Figure 8.39.

$$\beta = f(q) \tag{8.5}$$

A more detailed analysis of Foz do Areia data afforded a unique opportunity to study the effect of strangulation variation of the independent parameter D/h (the measurements

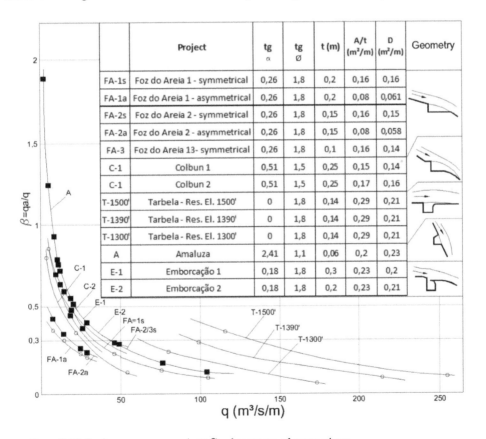

	Project	tg α	tg Ø	t (m)	A/t (m²/m)	D (m²/m)	Geometry
FA-1s	Foz do Areia 1 - symmetrical	0,26	1,8	0,2	0,16	0,16	
FA-1a	Foz do Areia 1 - asymmetrical	0,26	1,8	0,2	0,08	0,061	
FA-2s	Foz do Areia 2 - symmetrical	0,26	1,8	0,15	0,16	0,15	
FA-2a	Foz do Areia 2 - asymmetrical	0,26	1,8	0,15	0,08	0,058	
FA-3	Foz do Areia 13- symmetrical	0,26	1,8	0,1	0,16	0,14	
C-1	Colbun 1	0,51	1,5	0,25	0,15	0,14	
C-1	Colbun 2	0,51	1,5	0,25	0,17	0,16	
T-1500'	Tarbela - Res. El. 1500'	0	1,8	0,14	0,29	0,21	
T-1390'	Tarbela - Res. El. 1390'	0	1,8	0,14	0,29	0,21	
T-1300'	Tarbela - Res. El. 1300'	0	1,8	0,14	0,29	0,21	
A	Amaluza	2,41	1,1	0,06	0,2	0,23	
E-1	Emborcação 1	0,18	1,8	0,3	0,23	0,2	
E-2	Emborcação 2	0,18	1,8	0,2	0,23	0,21	

Figure 8.39 Performance curves ($q \times \beta$) of aerators of some plants.

*Translation:*PROJETO = ProjectGEOMETRIA = GeometrySimétrico = SymmetricalAssimétrico = AsymmetricRes. El. = Res. WLSource: (Pinto, 1989).

covered a wide range of q – ranged from 7 to 100 m³/s/m and an aeration tower was closed). Foz do Areia data were plotted on a graph of $\beta \times Fr$ (Figure 8.40).

Curves A and B were identified to illustrate the function $\beta + f(Fr)$ for the two different strangulation conditions of the aerator 1 for symmetrical and asymmetric conditions. Similar curves can be identified for aerators 2 and 3. Note the high rate of variation of β when the number of Fr increases to an apparent limit.

To represent the prototype data analytically, ignoring the differences in geometry of the aerators, a simple function of the type $\beta = a(Fr - K)^b \, x(t/h)^c \, X(D/h)^d$ was adjusted by the least square method. The influence of t/h for being small was neglected and the following equation was found with a correlation factor of 99%.

$$\beta = 0,47(Fr - 4,5)^{0,59} \; x \; (D/h)^{0,60} \tag{8.6}$$

This equation is shown in Figure 8.40 for some values of D/h. The indicated values of D/h of each measured prototype data clearly confirmed the expression.

Following the good results of the analysis, an expression for general data was attempted, despite the diversity of the geometry of the aerators and the lack of a theoretical basis for the simple monomial function. It was concluded by the following formula, which adjusted well to the results measured with a correlation factor of 97,62%.

$$\beta = 0,29(Fr - 1)^{0,62}(D/h)^{0,59} \tag{8.7}$$

Pinto (1989) suggests that this expression can be used for an estimate of the amount of air entrained in the initial phase of a project. As mentioned earlier, an air concentration of 5 to 10% near the surface to be protected practically eliminates the risks of cavitation.

Reference should be made to the following works:

- "Cavitation in Chutes and Spillways," by Falvey H. T. Engineering Monograph no. 42 of the USBR (1990), available on the internet, which presents a procedure for the location of the aerators.
- *Development of Aerated Chute Flow*, de Kramer, K., thesis presented to the Federal Institute of Technology, Zurich, in 2004, to obtain a Doctor of Technical Science.
- *Analysis of Aeration in High Speed Flows in Spillways*, Brito, R. J. R., thesis presented to the School of Engineering of São Carlos, University of São Paulo, in 2011, to obtain a master's degree.

The theses of the last two authors mentions the difficulties to determine the minimum amount of air for the cavitation protection and the critical cavitation index, subjects that remain open. According to these authors, the need for further experimental studies on the evolution of air concentration that decreases along the flow remains evident.

As there is not yet a defined criterion to estimate this evolution and effective protection, based on the experience of Bratsk, Nurek, Foz do Areia, Emborcação and Guri, Pinto (1983) suggests that it seems reasonable to consider that a well-designed project an aerator should protect a section of chute from 50 to 100 m.

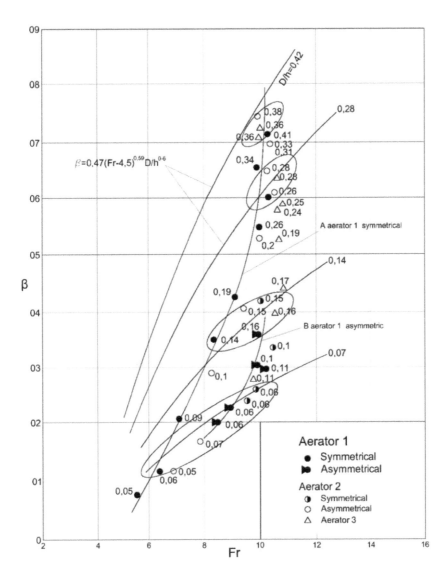

Figure 8.40 Curves *Fr* × *β* of aerator 1 of Foz do Areia.

Source: (Pinto, 1989).

8.6.2   *The case of Foz do Areia*

The Foz do Areia HPP (2.500 MW) was implanted on the Iguaçu river in the state of Paraná. Is a 160 m high concrete face rockfill dam.

The spillway, on the left abutment, with a capacity of 11.000 m³/s, is a classic profile with 4 sector gates of 14,5 × 18,5 m, followed by a chute with 70,6 m of width and 400 m in length, with a slope of 25,84% up to the flip bucket deflector, at an elevation 118,50 m below the maximum water level of the reservoir (Figures 8.41 and 8.42). The

Figure 8.41 Foz do Areia HPP. Chute spillway.
Source: (CBDB, 2002).

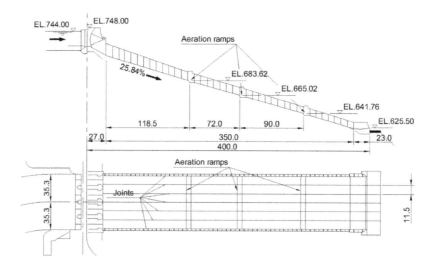

Figure 8.42 Foz do Areia HPP chute spillway. Plant and profile.
Source: (Pinto, 1983).

bucket velocity is of the order of 47 m/s, and the specific discharge is 155,8 m³/s/m. The main features and details of the designed aeration system are shown in the following illustration.

From Figure 8.42, it can be seen that the aerators were positioned 145,5 m from the spillway crest and 92,50 m from the jet takeoff lip. The three aerators were spaced 72 m and 90 m apart. According to Pinto, in 1981 the operation was adequate, but it was not possible to conclude whether two aerators would have been sufficient.

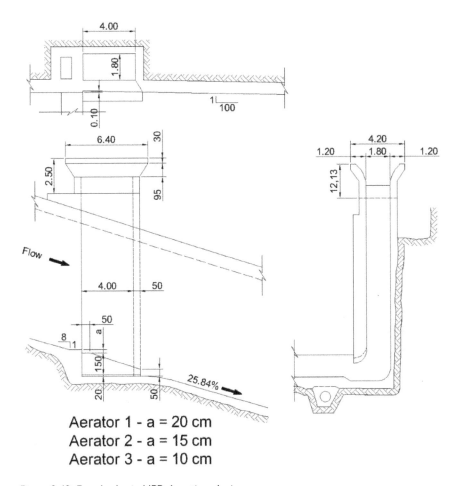

Aerator 1 - a = 20 cm
Aerator 2 - a = 15 cm
Aerator 3 - a = 10 cm

Figure 8.43 Foz do Areia HPP. Aeration devices.
Source: (Pinto, 1983).

8.6.3   *The case of Emborcação*

The Emborcação HPP (1.192 MW) in Paranaíba river, Minas Gerais state, is a project similar to that of Foz do Areia. The rockfill barrage with core is 158 m high.

The spillway, on the left abutment, with a capacity of 7.600 m³/s, consists of a classic sill, with four sector gates of 12,0 × 18,77 m. The chute is 58,5 m wide and 330 m long and has a slope of 18% up to the flip bucket deflector at an elevation 83,5 m below the maximum water level of the reservoir (Figures 8.44 and 8.45). The specific discharge is 130 m³/s/m. The flow velocity in the bucket is of the order of 40 m/s.

Figure 8.44A Emborcação HPP. Chute spillway.
Source: (CBDB, 2002).

Figure 8.44B Emborcação HPP. Side view.
Source: (CBDB, 2002).

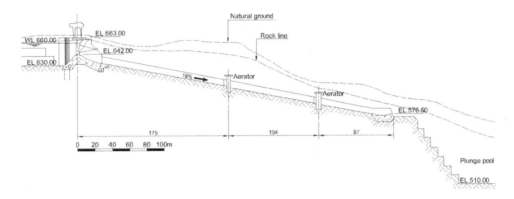

Figure 8.45 Emborcação HPP. Chute spillway profile.
Source: (CBDB, 2000).

Two aerators were provided spaced 103 m, the second being 65 m upstream of the jet takeoff lip, approximately.

8.6.4 The case of Xingó

Special mention should be made of the 3.000 MW Xingó HPP spillway project (Figures 8.46 to 8.50), which began operating in December 1994 (CBDB, 2002).

The spillway has a capacity of 33.000 m^3/s, with 12 gates of 14,8 m × 20,7 m in two chutes: the one on the right, all lined with concrete, is the service trough and has two aerators; the left channel, with a long unlined chute, only operates when the flow rates exceed 17.500 m^3/s (TR = 200 years). The chutes are 251 m long and 109 m wide each. The flow velocity in the bucket is on the order of 44 m/s, and the specific discharge in the right chute is 160,55 m^3/s/m. The structure is founded on a rocky gneiss massif sound of excellent quality.

Note: an economical solution; see also the cases already presented: Serra da Mesa (Figures 2.25 and 2.26) and Nova Ponte (Figures 5.73 and 5.74).

Figure 8.46 Xingó HPP. Chute spillways.
Source: (CBDB, Highlights, 2006).

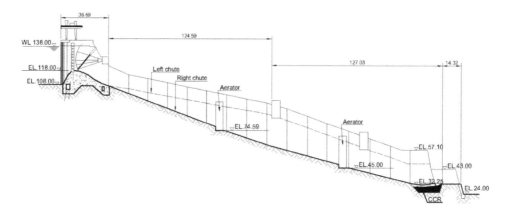

Figure 8.47 Xingó HPP. Spillway profile.
Source: (CBDB, 2002).

Figure 8.48 Xingó HPP chute spillways – right chute
Source: (CBDB, 2002).

It is worth noting that, despite not being directly linked to aeration, several erosion tests were carried out on the prototype in the unlined chute of this spillway in December 1995, in order to confirm the suitability of the adopted solution.

In the first test, on November 12, 1995, the discharge of 1.200 m³/s was passed during five hours. Erosions were observed.

In the second test, on December 12, 1995, the discharge was 2.400 m³/s for seven hours. Significant erosion was observed along the fault crossing the spillway at a 45° angle at the top of the chute. Erosion was along the foot of the central wall dividing the chutes. Downstream, at the end of the chute, the behavior was good.

In the third test, on December 14, 1995, the flow of 4.000 m³/s was passed for 30 minutes. Between the beginning and the end of this test, the operation lasted two hours. All unstable blocks, as well as dental concrete, were removed and a large depression

Figure 8.49 Xingó HPP. View of the left chute test – Q = 4,000 m³/s. This left chute will only operate at flow rates > 17,500 m³/s – TR > 200 years.

Source: (CBDB, 2002).

Figure 8.50 View after the operation test of the left channel of the spillway, with Q = 4.000 m³/s.

Source: (CBDB, 2002).

was formed between the two main discontinuities (Figure 8.50). There was no regressive erosion near the concrete lining. Erosion near the central wall remained as in the previous test. In some points it reached 1,0 m deep.

The behavior of the chute was considered adequate. Further treatment of the chute reinforcement and dental concrete was scheduled in a few places along the central wall to minimize future maintenance work.

The operation of the spillway was totally liberated, with the recommendation to maintain the erosion monitoring in cases of future spills with higher flows. The reference CBDB/LBS (2002), Large Brazilian Spillways, records that the spillway had not operated since 1998.

Chapter 9

Gates and valves

This chapter deals with the gates and valves used in spillways and in bottom outlets. For details on the gates projects, it is recommended to consult Erbisti (2004). It is recommended to consult also Kohler (1969) for the selection of gates and valves of the outlet works.

The construction of gates originated in the works of irrigation, water supply and inland navigation. In antiquity, the water was dammed by small dams and diverted to irrigation channels. The excess water was discharged over the dam. Then mobile dams were built, that is to say, dams with gates, which could be moved, opened, to give way to excess water, allowing greater flexibility of operation to the hydraulic works. With the exception of surface adduction channels (or low pressures pipes), pressurized adduction circuits normally use sluice gates and bottom valves to protect turbines and forced conduits.

Controlled spillways are normally provided with gates, including the bottom outlets (see Sobradinho HPP – Figures 2.27 and 2.29). Prior to selection, the design head and design discharge must be defined to determine the dimensions. The outlet dimension is the starting point for the selection of gates and valves, since the various types have maximum head limitation (maximum fall). The head limitation may be due to seals or possible hydraulic problems (cavitation tendencies and their damage, head losses, required maintenance and cost). Size is usually limited by manufacturing and transportation problems.

Even if the turbines have their own closing devices (distributor), it is prudent to envisage another mechanism for closing the circuit. The USBR has as a rule to provide at least two separate conduits, each one with a gate or guard valve.

This arrangement, although more expensive, ensures a high degree of freedom to control discharges. In addition, it enables maintenance and repairs to be carried out without stopping the operation, when the gate stop-flow becomes inoperative (which is what happens most often).

The conditions under which the conduits will discharge must also be established in the process of selecting the gate or valve. These conditions include determining whether the discharge will be made in the air or whether it will be submerged, when a basin of dissipation may be required. In addition, the degree of spray that the adjacent electrical installations can tolerate must be determined. For designs of gates and valves with air discharges, which are less hydraulically complex, it is necessary to be sure that they will never be submerged when operating the plant (spills and turbinated discharge increase the downstream level, and if there is a key curve error, may drown the conduit outlet). The hydraulic conditions of the project and topographical location of the plant may imply a solution with a dissipation basin. This fact does not strictly limit the type of gate or valve that can be used. But jet-gates, fixed-cone valves, and tube valves are not normally used.

In many cases, the discharge duct of a dam is a key aspect for a safe operation of the structure. For this reason, gates and valves for these conduits must be selected and arranged to operate hydraulically in a safe and viable way. Safety is paramount and cannot be compromised by the economy.

Many gate and valve designs have been made over the years. The rugged and simple are usually the best. Some commonly used gates and valves with proven safety are shown in the figures in Tables 9.3 and 9.5. In these tables, a brief summary of the characteristics of each of them is given by the type of function to be performed in the outletwork: discharge control (throttlinggates or valves) or guard (guard gates or valves).

Gates and valves with high flow coefficients and hydraulic efficiencies deserve the first consideration, but selection must be made considering all other factors listed. The terminology for describing the various types of closure devices was extracted from Davis (1952):

- Gate: a closing and opening device composed of a tray, with facing and framing, that moves through the flow of an external position to control the flow. Contains fixed parts and maneuvering mechanism. The seals, fixed to the face, usually consist of rubber profiles.
- Valve: a closing device in which the element of the action remains fixed axially with respect to the flow path and is rotated or moved longitudinally to control the flow.
- Gates or guard valves: These devices operate fully open or closed and act as a secondary safety device to cut off the flow if the primary flow closure device becomes inoperative. Guard gates are usually operated under pressure-balanced, non-flowing conditions, with the exception of emergency closures. Regulating gates and valves can be used to control the outflow of the reservoir or to control transient pressures occurring during turbine operation.
- Gates or throttling: These devices operate under high pressure to throttle the flow and control the rates of discharges, which are discharged freely, or at relatively low back-pressures. This does not include high back pressure in the strangulation. They can also control transient pressures during turbine operation.
- Stoplogs or cofferdam gates: usually installed at the entrance and used for securing the hydraulic circuit for inspection or maintenance and are always operated in a pressure-balanced manner. They cannot be considered as guard gates.
- Stoplogs: identical to cofferdam locks; a stoplog can be considered as a section of a cofferdam, which is made of several elements for ease of handling.

9.1   Types and components of the gates

There are several types of gates: sluice, hinged, cylindrical, stoplog, caterpillar, miter, rolling, segment, sector, Stoney, drum, roof and fixed roller gate. In Brazil, these types are defined in Brazilian Standard – Hydraulic Gates – Terminology, NBR 7259 (1982), as well as in the book *Design of Hydraulic Gates* (Erbisti, 2004). Some of them are shown in the illustrations in this book (Figures 9.1 to 9.4).

For details on the location of the sill beam, the position of the trunnion in the case of radial gates, as well as the stresses on the gates, beyond this reference, see HDS (1965).

The sill beam affects the height of the gate, the local pressures and, to a lesser extent without influencing the design, the flow coefficients for partial openings. Locating the sill at the axis of the dam results in a gate with the smallest height. When the sill is positioned upstream of the axis the jet under the gate tends to jump away from the contour of the structure, resulting in negative pressures that can cause cavitation damage to the crest. When the sill is positioned downstream of the crest the jet is directed downwards and tends to follow the profile of the crest, generally resulting in atmospheric or positive pressures immediately downstream of the gate.

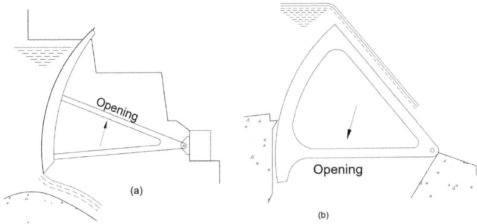

Figure 9.1 Gates: (A) Segment. (B) Sector.
Source: (Erbisti, 2004).

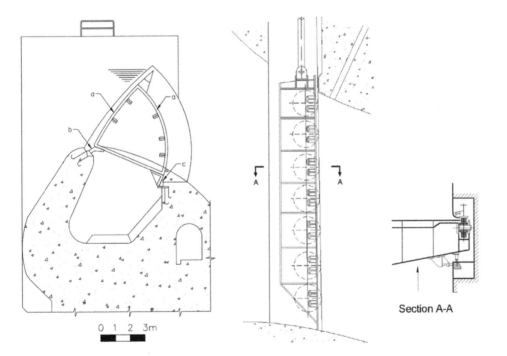

Figure 9.2 Drum gate.
Source: (Schreiber, 1978)

Figure 9.3 Fixed-wheel gate.
Source: (NBR 7529–1982).

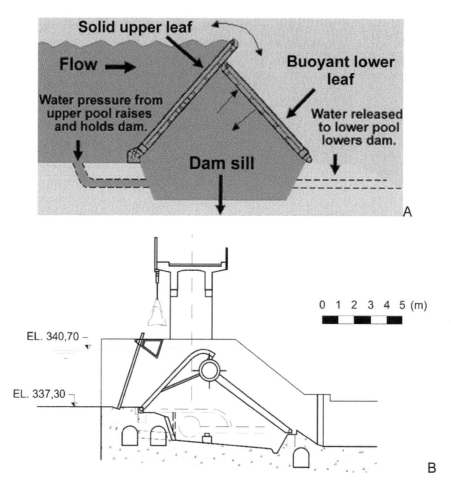

Figure 9.4 Bear-trap gate.

Source: (industrialscenery.blogspot.com).

The sill beam also affects the location of the trunnion position and height that the gate has to be raised to pass the maximum discharge (without gate control). According to HDS (1965), usually the sill beam is located 1,5 to 3,0 m away from the crest of the spillway.

The waterline profile can be determined with the aid of Figure 2.15. Normally, the position of the trunnion is above the upper waterline. The vertical position of the trunnion is approximately one-third the height of the gate. The location of the trunnion of the gate depends on the position of the gate and the radius of the gate. It's a structural issue.

The gate radius has been adopted, usually, with values between 0,73 and 1,10 H_d. Of course, the length of the pillar is directly related to the length of the trunnion.

It is necessary to mention the fusegates, designed in 1989 by **François** Lempérière (Figures 9.5 and 9.6). A simple, robust and safe system to increase storage and daily generation capacity as well as spill capacity. The system was patented by Hydroplus International. Detailed information can be found on the site of the company.

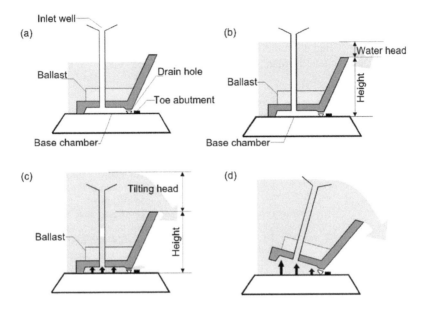

a) The water level is increasing;
b) Fusegate acts as an ungated spillway;
c) The water head reachs to the predetermined elevation (design tipping head); and
d) The fusegate tips.

Figure 9.5 Fusegate.

Source: (Alla, 1996; Abbas and Zeinab, 2012).

Figure 9.6 Fuse gate – Shongweni dam. South Africa (Hydroplus). Spillway width = 126,2 m' New capacity = 5.000 m³/s; number of fusegates = 10. height = 6,5 m; width = 9,73 m; increased discharge capacity = 235%.

9.2   Classification of the gates

The gates can be classified as function, movement, discharge, composition of the gate leaf, location and shape of the skin plate, as summarized in Table 9.1.

Table 9.1 Classification of the gates (Erbisti, 2004. Revision 12/01/2019).

Classification	Type	Utilization
Function	Service	Service gates are used for continuous regulation of flow or water level. Examples: spillway sluices; bottom outlet gates; lock gates (navigation chambers and aqueducts); flood control automatic gates.
	Emergency	Emergency gates are used occasionally to shut down the flow of water in conduits and channels; as a rule, are designed for normal operation in open or closed position. Examples: intake gates; gates installed upstream of penstock service valves; draft tube gates of Kaplan turbines; gates installed upstream of bottom outlet gates.
	Maintenance	Stoplogs are operated only with standing water and their main function is to allow the emptying of the conduit or channel for proper access and maintenance of the main equipment.
Gate leaf movement	Translation	Sliding: slide, stoplog and cylindrical. Rolling: fixed-wheel, caterpillar and Stoney.
	Rotation	Flap, miter, radial, segment, sector, drum, bear-trap and visor.
	Translo-rotation	Roller gate is the only type that performs a combined translation-rotation movement.
Discharge of the water relative to the gate leaf position	Over the gate leaf	Flap, sector, bear-trap and drum gates: in the opening operation they move downwards around the hinge axis situated on the sill, allowing the water passage over the gate leaf.
	Under the gate leaf	Slide, caterpillar, roller, segment, fixed-wheel, visor and Stoney gates move upwards, allowing the flow of water under the gate.
	Over and under the gate leaf	The mixed and double gates allow the discharge either over and under the leaf, depending on the operational requirements.
Composition of gate leaf	Plain	Gate leaf is composed of a single element.
	Mixed	The main gate leaf is provided at its top with a flap gate. Many applications are known of segment, fixed-wheel, roller and Stoney gates combined with flap gates, mainly in Europe.
	Double	The gate leaf is composed of two movable overlapping elements. By lowering the upper element, it is possible to discharge from above and vice versa. Both elements can be raised for maximum flood passage. Fixed-wheel and segment gates are the only known types of double-leaf gates.

(Continued)

Table 9.1 (Continued)

Classification	Type	Utilization
Location	Weir Submerged	All types of gates can be used in surface works. Only some types are applied in submerged installations: fixed-wheel, segment, caterpillar, slide, stoplog, cylindrical and Stoney. According to the water head over the sill, gates are usually classified as: • Low head gates: up to 15 m; • Medium head gates: from 15 to 30 m; • High head gates: over 30 m. This criterion is subjective and changes according to the technological evolution.
Shape of the skin plate	Flat	Slide, caterpillar, fixed-wheel, Stoney, stoplog and bear-trap.
	Radial	Segment, sector, drum, visor, cylinder and roller. Flap and miter may have a flat or curved skin plate. Reversed segment gates, which are very common in Germany, most have a flat skin plate.

A gate consists of three basics elements: gate leaf, fixed parts and maneuvering mechanism. Figure 9.7 shows the components of a fixed-wheel gate.

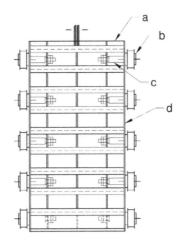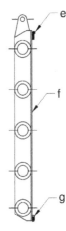

Figure 9.7 Fixed-wheel gate nomenclature: (a) gate leaf; (b) wheel; (c) wheel pin; (d) end girder; (e) top seal; (f) skin plate; (g) bottom seal.

Source: (Erbisti, 2004).

9.3   Selecting the type of gate

The choice should be based on an analysis of all factors that can influence the performance, cost, quality and reliability of the equipment, such as: operational safety; lower delivery weight; simplicity of operation; ease of maintenance; structural requirements (slots, chambers, guides, etc.); magnitude and direction of the forces transmitted to the concrete; maneuverability; ease of transportation and assembly. There are many factors. In the selection of the gate, it is also considered the experience acquired in projects carried out, as well as the experience of the manufacturers. Table 9.2 shows the applicable types of gates by structure/purpose.

Table 9.2 Applicable types by structure/purpose (Erbisti, 2004)

Structure	Type of Gate
Intakes	Fixed-wheel, slide, Caterpillar, segment and cylinder.
Spillways	Segment (compression or traction), flap, fixed-wheel, sector, drum, segment with flap, fixed-wheel with flap and double-leaf fixed-wheel hook-type.
Bottom outlets	Slide, fixed-wheel, Caterpillar and segment.
Lock gates	Miter, fixed-wheel and segment (with horizontal or vertical pivot axis)
Lock aqueducts	Slide, fixed-wheel and reversed-type segment.

9.4   Limits of use

The continuous development of hydroelectric power plants has considerably increased the application limits of each gate type (width, height and hydrostatic head). Examples:

- Itaipu HPP: 14 segment gates were installed with 20 m of width and 21,34 m of height;
- Tucuruí HPP: 23 segment gates were installed with of 20 m of width and 21 m in height.

The practical limits of use of the various types of gates can be found in Erbisti (2004). It is recommended to consult: Lemos de Lucca, Y. de F. *Hydromechanical Equipment*. São Paulo (2016), available in http://progestao.ana.gov.br/portal/progestao/destaque-superior/ boas-praticas/curso-de-seguranca-de-barragens-daee-1/parte-7-comportas.pdf, and Amorim, J. C. C. *Module 1: Dams. Legal, Technical and Socio-Environmental Aspects, Unit 7: Hydromechanical Aspects. Course on Dams Safety*. Org.: Itaipu Binational, ANA – National Water Agency and MME – Ministry of Mines and Energy, Brazil (2012).

9.5   Flow coefficients of outletworks

9.5.1   *Flat conduit gates*

According to Kohler (1969), the coefficients of discharge of the constrictions of the hydraulic circuits vary between 0.6 and 0.8; for guard gates, range from 0.9 to 1.0. These coefficients vary according to the specificities of each design, since they depend on the characteristics of the flow lines approaching and leaving the opening, which depend on the shape of the entrance and the position of the trunnion (Figure 9.8). See also Table 9.4 and 9.5 at the end of the chapter.

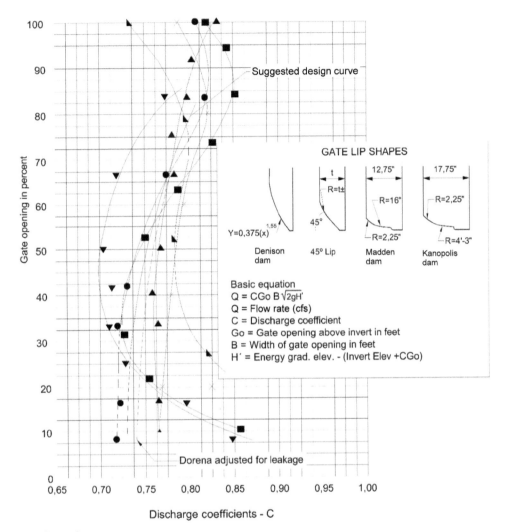

Figure 9.8 Flow coefficients applicable for partial openings of flat gates, fixed-wheel gates in conduits.

Source: (HDC, 1959).

9.5.2 Segment gates on bottom conduits

For the discharge coefficients of bottom conduits sector gates see Figure 9.9 (HDC 320–1). The curves shown are based on the R. von Mises equation obtained from the Garrisson tunnel model. For details see HDC (1959) and USBR – DSD (1974).

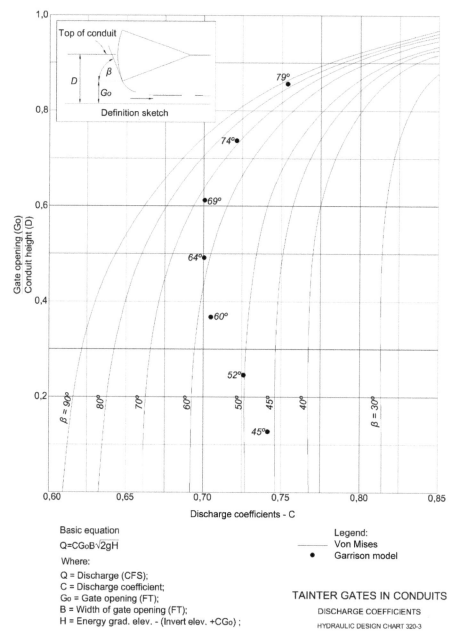

Basic equation

$Q = CG_0 B\sqrt{2gH}$

Where:

Q = Discharge (CFS);
C = Discharge coefficient;
G_0 = Gate opening (FT);
B = Width of gate opening (FT);
H = Energy grad. elev. - (Invert elev. +CG_0) ;

Legend:
——— Von Mises
● Garrison model

TAINTER GATES IN CONDUITS

DISCHARGE COEFFICIENTS

HYDRAULIC DESIGN CHART 320-3

Figure 9.9 Flow coefficients applicable for partial openings for segment gates on bottom conduits.

Source: (HDC, 1959).

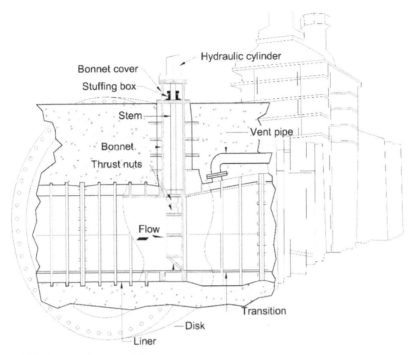

Figure 9.10 Bonneted gate – slide gate.
Source: (Rodney Hunt).

Figure 9.11 Half of a bonneted gate. Assembly – slide gate.
Source: (Rodney Hunt).

Figure 9.12 Bonneted gate.

The bonneted sliding gates are usually used to regulate the flow through the outletworks of dams. They are completely encapsulated and are designed and manufactured to be embedded in concrete (except the maneuvering equipment). It is similar to a rectangular/square gate, or to a valve. Frequently they are used in conjunction with a downstream operating gate and an upstream emergency closing gate. They have been manufactured in maximum sizes of 3 m × 3 m and have been used for high falls (greater than 150 m – reference: Rodney Hunt).

Figure 9.13 shows the head loss coefficients for partially open circular and rectangular gates (Miller, 1978). For other gates see HDC (1959) available on the internet.

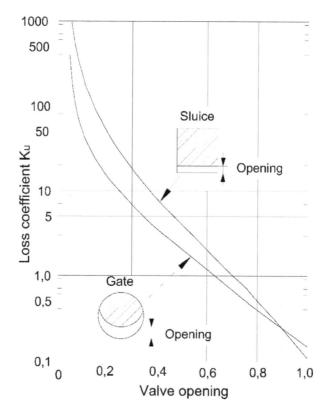

Figure 9.13 Head loss coefficients for rectangular and circular gates partially open.
Source: (Miller, 1978).

9.6   Discharge coefficients of spillways radial gates

The flow coefficient depends on the characteristics of the flow lines approaching and leaving the orifice, which depend on the shape of the crest, the beam radius and the position of the trunnion.

Figure 9.14 (chart HDC 311–1) presents a series of model and prototype data for various crest shapes and radial gate designs for non-submerged flows (see also Figures 9.15 to 9.20A).

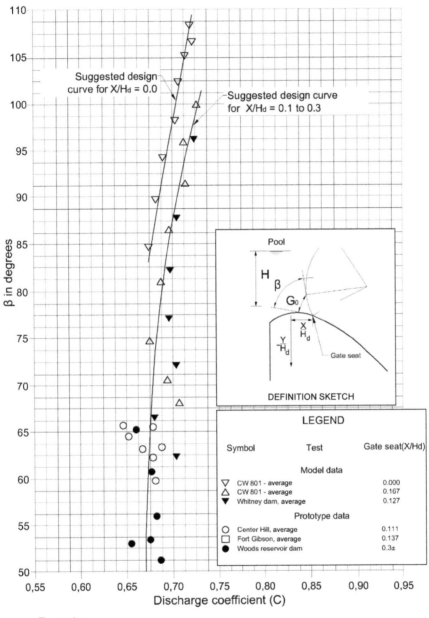

Formula

$$Q = CGo\ B\sqrt{2gH}$$

Where:
Go = Net gate opening
B = Gate width
H = Head to center of gate opening

TAINTER GATES ON SPILLWAY CRESTS
DISCHARGE COEFFICIENTS
HYDRAULIC DESIGN CHART 311-1

Figure 9.14 Flow coefficients for sector (radial) surface gates.
Source: (HDC, 1959).

Figure 9.15 Itaipu HPP spillway; 14 segment gates: 20,00 m × 21,34 m.

Figure 9.16 Balbina HPP. 5 Segment gates: 13,50 m × 15,30 m.

Figure 9.17 Tucuruí HPP spillway; 23 Segment gates: 20,00 m × 20,75 m.
Source: (Eletronorte).

Figure 9.18 Segment gates of Barra do Braúna HPP. Dimensions 13 m × 16 m.
Source: (Denge).

Figure 9.19 Slide gate São Bernardo HPP. Dimensions 2,50 m × 5,62 m.
Source: (Denge).

Figure 9.20 Flap gates.

Figure 9.20A Flap gate Juba II HPP. Dimensions 75,00 m × 1,00 m.

Source: (Denge).

The data presented are mainly based on tests with more than three spans in operation. The discharge coefficients for a single span shall be smaller because of lateral contractions. At the time, they had no data available to present. For details, it is recommended to consult HDC (1959).

For further information on the other gate design items (efforts and requests, maneuvers, drive systems, construction materials, fences, fabrication, transportation and assembly) see Erbisti (2004).

9.7   Risks of the gates

The risk that a spillway gate fails to open is a significant factor in dam rupture around the world. If the gates fail to open during a flood the dam can be overtopped and fail. This fault is responsible for 20 to 30% of dam ruptures, according to Leyland (2008).

Experience shows that significant improvement in gate safety is needed to reduce this risk. In many cases the gates are unable to operate due to failure of the control system or in the power supply system during the flood. In some cases, the gates fail because operators fail to start opening them. For details it is recommended to see Leyland (2008) and (2014a), among others.

In 2008, Leyland proposed, based on his experience, a new system to open the gate ("A new system for raising spillway gates"). The system provides for each gate a pair of hydraulic cylinders that push the gate instead of pulling. In other words, these cylinders extend to raise the gate. It provides independent back up for each gate, also driven by independent hydraulic system. It is much safer, cheaper than the long cylinders of the traditional system of pulling (tensioning) the gate, which, even, requires special manufacture.

In 2014, Leyland quoted Hinks and Charles (2004) who reported that in Norway during the flood of 1987, 50% of dam owners experienced problems with power outages, 23% had communications problems, 19% had problems with gates that did not open and 17% had problems with damage to the access roads to the plants.

Still according to Leyland (2014a), the Dam safety manuals from several countries specify that the spillway should be able to pass the flood of design with an out-of-operation gate. According to him some engineers seem to believe that if the prescriptions of the Manuals are followed the system of gates will automatically be safe, incredible as it may seem. It is emphasized that the safety of the gate can only be ensured if each individual part of the system has been carefully examined to determine failure modes and if steps are taken to ensure that the risk of rupture is extremely low.

According to Bowles (2007), the risk of dam breaking from an incident should be on the order of 1:10.000. Under many circumstances, failure of all gates will result in the dam overtopping and the consequent failure.

Micovic (2014) has posed that the greater danger is one in which several or all gates fail to open to pass moderate floods, instead of the rupture due to gate failure in an exceptional flood. The risk of all gates failing to open must be at least less than 1:1.000 instead of ~ 1:100 which is not usual for most installations.

Jansen (1983) reported that regular maintenance is essential and often receives little attention. Expensive, carefully designed equipment has often been neglected, especially where frequent operation is not required.

The malfunction of gates, valves, lifting equipment, can result from: displacement of the structure; corrosion and aging; misalignment of parts; insufficient lubrication; improper operating procedures; lack of energy; faults in the electrical circuits. Improper maintenance leading to failure of electrical and mechanical equipment can lead to dam failure at a crucial time.

Leyland (2014a) cited examples of Macchu 2 in India, Sayano in Siberia and the Kentucky dam in the United States of America.

In Macchu 2, on August 11, 1979, the dam broke and 2.000 people drowned (Jansen, 1983). Investigations showed that 3 of the 18 radial gates were jammed and could not be opened by the electric winch system, or by the manual backup system (Figures 9.21 and 9.22). A flow of 13.450 m^3/s passed through the spillway. The dam was overtopped 0,6 m and 700 m of the right dam and 1.070 m of the left dam was washed out. Unbelievably, the investigating committee was dissolved before its report was completed.

In Sayano, on August 17, 2009, the powerhouse was flooded due to the catastrophic rupture of one of the turbines and the power supply was lost to the lifting system of the 11 spillway gates. Fortunately, an emergency supply was provided and the gates were lifted before the reservoir reached a critical level. If the gates had not been opened on time, the dam would collapse and a million people would be at risk. There are now two cranes to move the gates. See also Leyland (2010). Figure 9.23 shows the plant after the repair work. Note the additional spillway on the right bank.

The Kentucky dam has a spillway with 24 spans designed for 30,000 m^3/s with two lifting equipment (Figure 9.24). In order to move the gates multiple claws have to be lowered into the flow to catch the gates, which is always very difficult, requiring a lot of skill from the operator. It takes a team of five operators.

The fact that there are two sets of lifting equipment does not add much to the overall safety because the two are parked in the powerhouse at the end of the spillway and if one fails the second cannot pass to open the gates.

Figure 9.21 Machhu 2 dam. View of the break.

Source: (USBR, Dam Failure Case Histories – An Introduction. Bruce Feinberg).

Figure 9.22 Machhu 2 dam. Detail of the break.
Source: (USBR, Jansen, 1983).

Figure 9.23 Sayano HPP. Downstream view after the repair work. Note the additional spillway on the right bank.
Source: (russiatrek.org or photo-day.ru).

As TVA is operating its power plants by remote control, it will be more difficult to ensure that a team of five people will be always available to open the gates in stormy weather, when the highways may be congested and their families in danger.

The Kentucky dam is not an isolated case. There are many others in the USA and the world in a similar situation.

It is recommended to consult the articles of Leyland (2008 and 2014a), available on the internet, which cites other examples, such as the West African dam in Africa.

It is also recommended to consult Hartung (1973), who discusses the hazards associated with spillway gates and the characteristics of the various types of gates in connection with the spillways of large dams.

 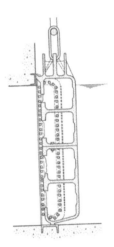

Figure 9.24 Kentucky dam. Claw detail of the caterpillar gate.

Source: (commons.wikimedia.org).

9.8   Rupture of radial gate of the Folsom dam spillway

It is reported that, although infrequent, the dam at the Folsom dam spillway in California, July 17, 1995, was broken, which had been in operation by the USBR since 1956 without any sign of structural danger. No one was injured in the accident. The flow through the Lower American river was 1.132 m³/s.

The gates have the following characteristics: 15,5 m high, 12,8 m wide, 14,33 m radius, weight of 87 t.

The investigations revealed that a diagonal joint in the arm of the dam, adjacent to the trunnion, was the initial point of rupture (Todd, 1999).

According to Anami *et al.* (2013), the gate broke because of flow-induced vibrations in operations with small openings.

Figures 9.25 through 9.29 show some photos of the incident and recovery work, as well as the construction of the additional auxiliary spillway.

In Appendix 1, there are cases of spillway rehabilitation, including another case of segment gate break – La Villita dam, Mexico, on May 31, 2013.

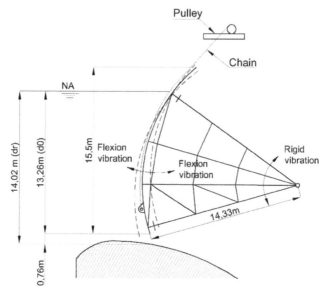

Pulley

Chain

NA

14,02 m (dr)

13,26m (d0)

15.5m

0.76m

Flexion vibration

Flexion vibration

Rigid vibration

14,33m

Figure 9.25 Gate section
Source: (Anami, 2013).

Figure 9.26 Gate breached.
Source: (www.flickr.com).

Figure 9.27 Gate breached – detail.
Source: (www.pbs.org).

Figure 9.28 Installation of new gate.
Source: (www.flickr.com).

Figure 9.29 Folsom dam. Observe the work of the auxiliary spillway close to the highway.
Source: (www.army.mil).

9.9 Valves

The valves are briefly described here. Needle valves are used in the outlet of conduits to control discharges under extremely high loads. They are designed to discharge into the air, thereby reducing the opportunity for cavitation inside the conduit. Although they have been used in a number of dams, for example Hoover dam, they have been supplanted by more economical and more efficient valves, such as fixed-cone valves and hollow-jet valves.

Fixed-cone valves, also known as Howell-Bunger, are widely used as flow control valves. Additionally, they correspond to a very good energy sink and aerator.

The hollow-jet valve is essentially half a needle valve, with the needle turned in such a way that it moves upstream in the closure.

Butterfly valves (Figure 9.30) are usually the most used as emergency valves upstream from hydroelectric plant generating units, where the penstocks are long, or as emergency shut-off valves in outletworks.

The spherical valve (Figure 9.31) consists of a large sphere in a housing with a cylindrical bore the same size as the conduit. When the valve is open, the cylindrical bore is aligned with the conduit. Turning 90°, the ball closes the valve. Head losses on a ball valve are insignificant, while costs are typically higher than those on a butterfly valve.

The two dominant valves can be compared as shown in Table 9.3. For more details see Table 9.6. It is also recommended to consult Kohler (1969).

Finally, the example of the solution adopted at the Funil Plant, in Paraíba river, Rio de Janeiro state, Brazil, shown in Figure 9.32. The plant has two tunnel spillways, one on each bank, and a Howell-Bunger spreading valve to control flow and energy dissipation (Figures 9.32 and 9.33) with $d = 3,50$ m. The flow varies between 190 m^3/s $< Q <$ 285 m^3/s.

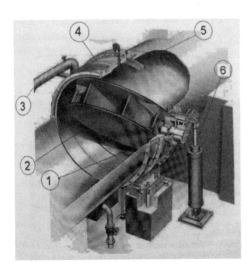

Figure 9.30 Butterfly valve1) Valve housing. 2) Biplane disc (patella). 3) Pipe bypass. 4) Valve housing with main seal. 5) Mobile seal assembly. 6) Servomotor.

Source: (Vinogg and Elstad, 2003).

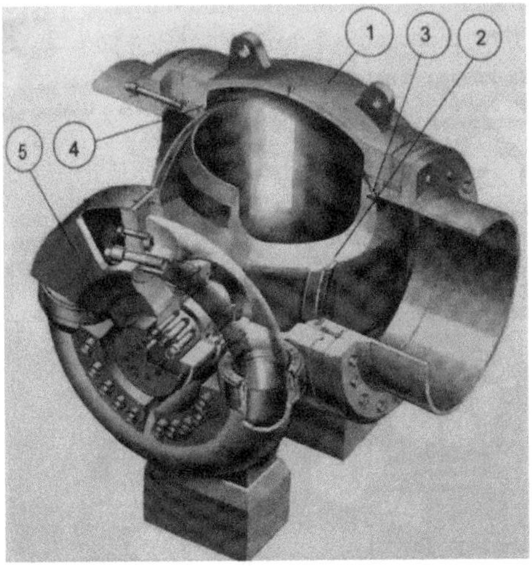

Figure 9.31 Spherical valve.1) Trunnion assembly. 2) Valve rotor. 3) Maintenance seal. 4) Main seal. 5) Ring piston servomotor.

Source: (Vinogg and Elstad, 2003).

Table 9.3 Comparison of guard valves (Vinogg and Elstad, 2003).

	Butterfly valve (Figure 9.30)	Spherical valve (Figure 9.31)
Head H (m)	Over 200–300 m	For higher heads
Diameter D (m)	Over 5,0–6,0 m	Over 3,0–4,0 m
Head loss coefficients	0,2–0,5	0,02–0,05
Flow coefficients (*)	0,7–0,8	1,0

(*) The coefficients are approximate and may vary depending on the characteristics of the projects. It is recommended to consult additionally HDC (1959) – Chart 330.

Figure 9.32 Funil HPP.

Figure 9.33 Funil HPP. Valve running.

It should be noted that the butterfly valves (Figures 9.34) are most often used as a guard valve, upstream of the turbines, where the penstocks are long. They can also be used as a dispersing valve (Figure 9.35), for emergency closure of the discharge conduit. Figure 9.36 shows a spherical valve.

A   B

Figure 9.34 Butterfly valve: A) closed; B) open in assembly.

Figure 9.35 Dispersing valve, type hollow-jet, or Howell-Bunger/fixed-cone.

Source: (kipab-co.com).

Figure 9.36 Spherical valve.

Source: (www.zeco.it).

Table 9.4A* Throttling Gates (Kohler, 1969)

Service Classification	Throttling gates				
Schematic Diagram					
Name	Unbonneted Slide Gate	Bonneted Slide Gates		Jet-Flow Gate	Top Seal Radial Gate
		High Pressure Type	Streamlined Type		
Max. Head (approx.)	75'	200'	500'	500'+	200'- 500'
Disch. Coefficient (a)	0,6 to 0,8	0,95	0,97	0,80 to 0,84	0,95
Submerged Operation	No	No	Yes (1)	Yes (1)	No
Throttling Limitations	Avoid Very Small Disch.	Avoid Very Small Disch.	Avoid Very Small Disch.	None	None
Spray	Minimum	Minimum	Minimum	Small	Minimum
Leakage	Small	Small	Small	None	Small to Moderate
Nom. Size Range (b)	To 8' wide & 12' high	To 8' wide & 9' high	To 10' wide & 20' high	10' to 20' dia.	To 15' wide & 30' high
Availability	Commercial Std. (1)	Special Design	Special Design	Special Design	Special Design
Maintenance Required	Paint	Paint	Paint (1)	Paint	Paint – Seals (1)
Comments and notes (a) Coefficients are approximate and may vary somewhat with specific designs (b) Size ranges shown are representative and are not limiting	(1) Gates are readily available from several commercial sources. They are not on off-the-shelf item, however		(1) Air vents required (2) Use of stainless steel surfaced fluidways will reduce pointing requirements and cavitation damage hazard	(1) Air vents required	(1) Seal replacement in 5–15 years is probable depending on design and use

(*) Original table, no update.

Table 9.5B* Throttling Valves (Kohler, 1969)

Service Classification	Throttling Valves				
Schematic Diagram					
Name	Fixe-Cone Valve	Hollow-Jet Valve	Needle Valve	Tube Valve	Sleeve Valve
Max. Head (approx.)	1000'	1000'	1000'	300	
Disch. Coefficient (a)	'0,85	0.70	0,45 to 0,60	0,50 to 0,55	0,80
Submerged Operation	Yes (1)	No (1)	No	Yes	Yes (1)
Throttling Limitations	None	Avoid very small disch.	None	None	None
Spray	Very Heavy (2)	Moderate	Small	Moderate (1)	None
Leakage	None	None	None	None	None
Nom. Size Range (b)	6" to 108' dia.	30" to 108' dia.	10" to 96' dia.	36" to 96' dia.	12" to 24" + dia. (2)
Availability	Commercial St. (3) Paint	Special Design Paint	Special Design Paint (1)	Special Design Paint	Special Design Paint
Maintenance Required					
Comments and Notes:	(1) Air vent required	(1) Submergence to axis of valve is permissible	(1) If water operation is used disassembly at 3 to 5 years interval for removing scale deposits is usually necessary	(1) Spray is heaviest at openings of less than 35%. At larger openings the rating would be better than moderate	(1) Valve is designed for use only in fully submerged conditions
(c) Coefficients are approximate and may vary somewhat with specific designs	(2) Spray rating will change to moderate if a downstream hood is added				(2) Larger sizes seem feasible and will probably be developed
(d) Size ranges shown are representative and are not limiting	(3) Valves are not stock items but standard commercial designs are available				

(*) Original table, no update.

Table 9.6* Guard Gates and Guard Valves (Kohler, 1969)

Service Classification	Guard Gates					Guard Valves	
Schematic Diagram	See Figure 9.3A for Diagrams						

| Name | Slide Gates | | | Ring-Follower Gate | Wheel-Mounted or Roller Mounted Gates | Butterfly Valve | Spherical and Plug Valves |
	Unbon	Hi-Pres	Strml				
Max. Head (approx.) (a)	100'	250'	500'	500'	500'	750'	1,500' +
Head Loss (1)	(1)			Negligible	Small to Moderate	None to Small (1)	Negligible
Leakage	*	*	*	None			None
Nom. Size Range (b)	*	*	*	36" to 120" dia.	To 10' wide & 30'high	12" to over 12'dia.	12" to over 10'dia.
Used as Guard Unit for	(2)		(2)	All types of circular conduit throttling gates and valves	Top seal radial gates and other square or rectangular units	All types of circular conduit throttling gates and valves	All types of circular conduit throttling gates and valves
Availability	*	*		Special Design	Special Design	Std and Special (2)	Std and Special (1)
Maintenance Required	*	*		Paint	Paint – Rubber Seals	Paint – Seals (3)	Paint
Comments and Notes: (e) Head losses are approximate and may vary somewhat with specific designs	(*) See data on Fig. 9.3A. (1) Head loss coefficients will vary from about 0.2 to 0.4 depending on entrance				Normally wheel-mounted gates are used except for high heads	(1) Rubber seated valves have no leakage when new. Metal seats will have some leakage. (2) Sizes to 36" or 18" are fairly standard. Larger sizes and high pressures are usually special.	(1) Sizes to about 24" are fairly standard. Larger sizes and high pressures are special

(Continued)

Table 9.6* (Continued)

Service Classification	Guard Gates	Guard Valves
(f) Size ranges shown are representative and are not limiting	(2) Usually used with a similar type throttling gate. Sometimes used for other types (3) Used close coupled with similar throttling gate. See Fig. 9.3A	(3) Metal seals may require periodic adjustment

(*) Original table, no update.

Chapter 10

Hydraulic models

10.1 Summary of theory

Reduced models are important for testing and optimizing the designs of permanent and temporary hydraulic structures. The following abstract was prepared on the basis of Rouse (1938, 1950, 2011), Davis (1952), USBR-HLT, Hydraulic Laboratory Techniques (1980) and Liggett (1994).

Until the end of the nineteenth century, hydraulics was basically empirical, to the extent that it was defined as the "science of coefficients." At the end of the eighteenth century and during the nineteenth century there was remarkable experimental work on flow in hydraulic conduits and in singularities (orifices, nozzles and spillways).

This brilliant phase of the hydro-experimental technique represents the beginning of modern hydrodynamics, and among dozens of others, Darcy, Bazin and Froude stand out. They, like the others, performed their experiments with limited resources and within restricted objectives, aiming to characterize the mechanism of hydraulic phenomena as a natural and general fact.

The hundreds of empirical expressions obtained at that time demonstrate a great amount of inquisitive efforts, which were also directed towards the abstract field of human knowledge, based on the definition of the utopian liquid (perfect liquid).

Great progress was made in the analysis of flow phenomena of flow through the application of the mathematics by Bernoulli, Euler and Lagrange.

As a decisive landmark in the conceptual evolution of hydrodynamic, mention should be made of the works of Osborne Reynolds, developed at the end of the nineteenth century, concerning the characterization of the flow, laminar and turbulent, and the attention of the German school, which, like Ludwig Prandtl and Von Karman, promoted a thorough review of hydraulic research methods, through important work on turbulence, giving remarkable impetus to aerodynamic and hydrodynamic in the early twentieth century.

These new concepts profoundly changed the course of hydraulics to the point that in the present day the experimental technique has its own characteristics, very different from those existing in the past.

Parallel to the evolutive aspects mentioned above, another notable occurrence was recorded, namely the use of two mathematical tools of work, the theories of dimensional analysis and mechanical similarity, discovered by Newton, which resurfaced as powerful levers and boosted the experimental technique in the field of hydraulics.

Dimensional analysis expanded with Buckingham's work (Pi Theorem), developed in 1914. Mechanical similarity became especially significant after the works of Reynolds (1883), Reech (1852), and others.

The combination of these two mathematical tools, working on the same occasion as a great outbreak of technological and industrial development, provided remarkable initiatives in the field of experimental technique through the use of the hydraulic laboratory.

The Germans, who at the time were developing important projects related to fluvial hydraulics, were the first to organize themselves.

In 1898, Hubert Engels, a professor at the University of Dresden, set up the first fluvial hydraulic laboratory to deal with the problems of fluvial flows according to hydraulic-experimental techniques similar to those used today.

Then, in 1901, Theodor Rehbock set up another laboratory at the Karlsruhe Institute of Technology. Since then, in the first half of the twentieth century, several other hydraulics laboratories have been set up around the world.

10.2 Definition of the model and laws governing the models

A model can be defined as a system through which and by its operation the characteristics of other similar systems can be predicted. Hydraulic models are generally smaller than their prototypes.

In general, the laws governing model-prototype relations are derived from the laws governing the action of the phenomenon under investigation. In cases where the law governing action on the prototype is unknown or very little known, the models can only be used qualitatively.

The results obtained from the hydraulic model can be transferred to the prototype by using laws of the model, which can be developed from the principles of dynamic similarity.

Similarity can be defined as a known and usually limited correspondence between the model and the prototype, with or without geometric similarity. This correspondence is rarely perfect, because it is generally impossible to satisfy all the conditions required for complete resemblance. However, these conditions are known as will be shown below.

The term similarity must therefore be qualified to indicate the general limits of correspondence, or one can speak of various kinds of similarity, each with a set of limitations.

The most commonly used relationships can be expressed as dimensionless groups by their numerical value that characterizes the flow under consideration. The deduction of the laws of the model assumes that for equal values of the dimensionless characteristics, the corresponding flows in the model and in the prototype are similar. This assumption has been sufficiently verified by experimental results.

The dimensionless groups most commonly used in hydraulic experiments are the Froude number, the Reynolds number and the Weber number. They are derived from the consideration of the forces of gravity, viscosity and surface tension, respectively, in conjunction with the inertial resistance force. The deduction of these laws can be found in Rouse (1938/ 2011), USBR-HLT (1980), Liggett (1994) and Chanson (1993), among others.

The law of Froude expresses to the similarity condition of the forces of gravity and inertia; that is, when the Froude numbers are equal in the model and the prototype, the relations of the forces of gravity and inertia are the same, and the patterns of the flows are similar.

$$Fr = \frac{V^2}{Lg} \tag{11.1}$$

This simple application of Froude's law assumes that neither viscosity nor surface tension influence the phenomenon under investigation. The Froude number is used as the basis for designing and interpreting models in which friction forces are negligible, such as in spillways and other hydraulic structures in which a rapid change in elevation of the water level occurs at a short distance, for example, as in the hydraulic jump in the stilling basins.

The Reynolds number (Equation 11.2) establishes the conditions under which the relations between the viscosity forces for the inertial forces are the same in the model and the prototype.

$$Re = \frac{VL\rho}{\mu} \tag{11.2}$$

Note: Several French researchers have used this relationship before:

1 Jean-Baptiste Bélanger was the first to introduce the relationship, in 1828, for channel flow: $V/\sqrt{gd}$; when $V/\sqrt{gd} < 1,0$ the flow is laminar; when greater than 1.0 the flow is torrential;
2 Dupuit used this same relationship in 1848;
3 Reech, F. (1805–1880), a shipbuilder, used the concept to test boats and propellers in 1852
4 Bresse in 1860 and Bazin in 1865;
5 William Froude, in the work of quantifying the resistance of floating objects, used a series of boat shapes to measure the strength of each one when towed at a given speed. The observations allowed Froude to derive, in 1868, the wave-line theory (WLT), which first described the resistance in a way as a function of the waves caused by the pressure variation around the hull as it moves through the water. In France, the number is sometimes called the Reech-Froude number (Chanson, 1995).

Equality of the Reynolds numbers in the model and prototype indicates that the flow patterns are similar when the viscosity and inertia forces are the forces governing the phenomenon. The forces of gravity and surface tension are negligible. The Reynolds number is mainly applied to closed flow systems, such as pipes or conduits where there is no water-free surface.

The Weber number (Equation 11.3) expresses the equality of the relations of the surface tension forces to the forces of inertia in the model and the prototype.

$$We = \frac{V^2 L\rho}{\sigma} \tag{11.3}$$

It is useful in certain problems of surface waves, formation of air bubbles, addition of air in water flows and other related phenomena. The use of Weber's number as a criterion assumes that the forces of gravity and viscosity are negligible.

The symbols mean:

* V = speed characteristic of the system; it can be the average velocity, the superficial velocity or the maximum velocity;
* L = linear characteristic or dimension, such as diameter or depth;
* g = acceleration of gravity;

- ρ = fluid density;
- μ = viscosity of the fluid;
- σ = surface tension of the fluid.

All these symbols need to be expressed in a consistent system of units such that the numbers are dimensionless. There are other dimensionless relationships in other fields, but they have no practical importance in the problems of river models and hydraulic structures. Dimensional criteria involving elasticity are important for the study of water hammer.

The design of models is not so simple that the application of one or another number of these is sufficient. Often two or three need to be used, and the decision process involves the specialist engineer. The difficulty found is illustrated by considering the speed variation with the length necessary to ensure similarity by one of these three criteria.

If all quantities remain the same, as in the case when the model fluid is water, each number requires that:

- the Froude number: $V \propto L^{1/2}$;
- the Reynolds number: $V \propto L^{-1}$;
- the Weber number: $V \propto L^{-1/2}$;

Two of these requirements cannot be satisfied simultaneously unless the model is on the 1:1 scale.

10.3   Types of models

The models can be divided in two types: models of structures and models of rivers. These two types are distinguished by the behavior of the water surface.

In the structural models there is a rapid change in the water surface and a corresponding dependence on the Froude number for the similarity. In river channel models, the change in elevation of the water surface is very gradual, being governed mainly by friction. In such models, similarity is governed by the Reynolds number or the laws of friction in river channels.

Both types sometimes occur in a single model, such as a sudden constriction in a river model caused by bridges or cofferdams. In such cases, it is necessary to satisfy both: the law of Froude and the law of friction at the same time, or if this is not possible, the model must be designed for the most important aspect.

10.3.1   Models of hydraulic structures

In the design of hydraulic structures models, the following considerations need to be made: (1) the desired results; (2) the available space for model construction; (3) the supply of water; and (4) the cost.

In general, the model must be constructed as large as possible in order to obtain more accurate results. Space constraints often command size, and in many cases, size is also limited by the availability of water and resources. The fundamental relationships for structural models in Froude's law are given in Table 10.1.

Choosing the length scale determines all others. Once the length scale is chosen, the size of the model and the water supply are determined. If the amount of water is limited, this determines the discharge scale, and hence the ratio of the length and size of the model.

Table 10.1 Flow characteristics and similarity relationships.
            Rouse (1961). USBR-HLT (1980).

Characteristic	Dimension	Scale Relationships	
		Froude	Reynolds
Geometric			
Length	L	L_r	L_r
Area	L^2	L_r^2	L_r^2
Volume	L^3	L_r^3	L_r^3
Kinematic			
Time	T	$\left(\frac{L\rho}{\gamma}\right)_r^{1/2}$	$\left(\frac{L^2\rho}{\mu}\right)_r$
Velocity	LT^{-1}	$\left(\frac{L\gamma}{\rho}\right)_r^{1/2}$	$\left(\frac{\mu}{L\rho}\right)_r$
Acceleration	LT^{-2}	$\left(\frac{\gamma}{\rho}\right)_r^{1/2}$	$\left(\frac{\mu^2}{\rho^2 L^3}\right)_r$
Discharge	L^3T^{-1}	$\left[L^{5/2}\left(\frac{\gamma}{\rho}\right)^{1/2}\right]_r$	$\left(\frac{\mu}{\rho}\right)_r$
Dynamics			
Mass	M	$(L^3\rho)_r$	$(L^3\rho)_r$
Force	MLT^{-2}	$(L^3\gamma)_r$	$\left(\frac{\mu^2}{\rho}\right)_r$
Pressure, intensity	$ML^{-1}T^{-2}$	$(L\gamma)_r$	$\left(\frac{\mu^2}{L^2\rho}\right)_r$
Impulse and momentum	MLT^{-1}	$[L^{7/2}(\rho\gamma)^{1/2}]_r$	$(L^2\mu)_r$
Energy and work	ML^2T^{-2}	$(L^4\gamma)_r$	$\left(\frac{\mu^2}{\rho}\right)_r$
Power	ML^2T^{-3}	$\left(\frac{L^{7/2}\gamma^{3/2}}{\rho^{1/2}}\right)_r$	$\left(\frac{\mu^3}{L\rho^2}\right)_r$

Note: When "g" is the same in the model and prototype $\gamma/\rho = 1$.

In Davis (1952) details can be found on different types of models and the respective tests for the phenomenon that one wishes to research. For example: discharge coefficients (spillways without pillar and with pillar); gates discharge coefficients; characteristics of stilling basins; special spillway structures such as side spillways, morning glory and siphons; cavitation studies; emergency gates subjected to high heads; water towers; fish ladders; chutes and drops.

In the case of discharge coefficients of the spillways, it is assumed that the model faithfully represents the action of the prototype. The discharge ratio, $Q = L^{5/2}$, assumes that the flow in the model and the prototype are similar and that the discharge coefficient in the model and the prototype are the same.

10.3.2   Structures in which friction is important

One can find in Davis (1952) details on the tests of structures in which friction plays an important role. For example: locks filling systems; spillways in tunnels; surge tanks. For these cases there are no established rules. The studies will depend on the conditions of each case.

10.3.3   Models of the river channels

The design of river channel models presents more difficulties than models governed only by Froude's law. In the flow of rivers friction plays an important role. The change in the surface of the water is very gradual, such that the forces of gravity play a minor role except at the places of section changes, where the flow becomes markedly nonuniform.

The source of many problems encountered with river models is the size of the area to be investigated. Space limitations are usually such that: when the scale is selected to enable the construction of an undistorted model for an available space, the depths of the flow are so small that laminar flow can occur. The laws that govern laminar flow are very different from those that control turbulent flow and how turbulent flow occurs in nature must also occur in the model. The use of an undistorted model usually results in waterline profiles so smooth that differences in elevation cannot be measured satisfactorily.

These objections can be resolved by distorting the model. Distortion can be done in several ways. The slope can be increased by tilting the model by arbitrarily changing the flow scale, using different horizontal and vertical scales, or by changing the roughness.

Scaling relationships for river models are usually based on Manning's formula (friction).

$$V = \frac{D^{7/6}}{L^{1/2}\eta} \tag{11.4}$$

Relationship of inclinations:

$$S = \frac{D}{L} \tag{11.5}$$

Relationship of discharge:

$$Q = \frac{L^{1/2}D^{13/6}}{\eta} \tag{11.6}$$

In certain cases where the forces of gravity and friction must be simultaneously satisfied, the necessary condition can be found by equating the velocity for the two conditions.

$$L = \frac{D^{7/6}}{L^{1/2}\eta} \tag{11.7}$$

Of which we have η:

$$\eta = \frac{D^{7/6}}{L} \tag{11.8}$$

For non-distorted models, the expression is reduced:

$$\eta = L^{1/6} \tag{11.9}$$

The dimensions of the model can be determined in a first approximation through the above relations. In addition, the flow must be turbulent. According to measurements made, the US WES established that the flow is turbulent if the following expression occurs:

$$\frac{VR}{v} \geq 4.000 \tag{11.10}$$

where:
V = average velocity (m/s);
R = hydraulic radius (m);
v = kinematic viscosity (m^2/s).

For water at 20 °C, the kinematic viscosity is 0,00001 ($1,01 \times 10^{-6}$). For this value, the turbulent flow is ensured if $VR \geq 0.04$. A value lower than this should be considered as sufficient reason to increase the model, or if this is not possible the results should be interpreted with extreme caution. Values smaller than 0,02 should not be used under any conditions because the flow will certainly be laminar.

A   Mobile bed models

The phenomenon involving the investigation of riverbed material movement is a type of model often used. This is the most difficult model to design and operate, but it is the most satisfactory result.

The laws that govern the movement of material from the riverbed are very complex and yet little and imperfectly known. The movement of the bed material depends on the tensile forces exerted by the flow. If the water weight component parallel to the channel of the river is equated to the resistance offered by the channel and assuming that the resistance is uniform throughout the wet perimeter, the tensile force can be expressed by the following equation.

$$T = \gamma RS \tag{11.11}$$

where:
T = tractive force (t/m^2);
γ = specific weight of water (t/m^3);
R = hydraulic radius (m);
S = slope of the water line.

The critical traction force, or the force at which the movement of the sand particles begins, for several researchers, is proportional to the grain diameter of the sand and is given by:

$$T_c = KD \tag{11.12}$$

where:
Tc = critical traction force (t/m^2);
D = diameter of the sand particle (m);
K = constant ranging from 0.2 to 0.4.

The traction forces in the models are extremely small and it is difficult to obtain the force necessary to cause the movement of the sand when the sand is used as bed material without a very strong slope of the water line. Attempts have been made to reduce the force required by using much lighter material bedding.

B   Fixed-bed models

These models, as the name says, are built with a fixed bottom (see Figure 10.1). They are used normally for:

- for detailed studies of the alternatives of the layouts of the works of the hydroelectric plants;
- for studies of optimization of the layout of the chosen alternative;
- for detailed studies of the river diversion (cofferdams);

A - Upstream view. Spillway access channel and intake structures.

B - Side view. In the foreground spillway and plunge pool pre-excavated.
In the background, intakes/powerhouses and tailwater channels.

Figure 10.1 Tucuruí HPP. Three-dimensional fixed-bed model. Final Report – 1982.Scale 1:150.
Source: HIDROESB (Rio de Janeiro – Brazil).

- or for investigations of the effects of permanent changes in the river channel caused by dredging, obstructions caused by bridge piers, or by dikes.

10.4   Materials and methods of construction. Equipment

For details on materials, constructive methods (curves or quoted planes) as well as on equipment see Davis (1952) and USBR-HLT (1980).

Figure 10.2 Jirau HPP. Construction of the three-dimensional model. Scale 1:110. Area = 1.000 m^2.
Source: (FURNAS Magazine, Year XXXI, n. 319, April 2005).

Figure 10.3 shows a photo of the mobile bottom model of Estreito HPP, on the Tocantins river (state of Pará, Brazil). See also Figure 6.46.

The following are photographs of reduced models extracted from the website of LACTEC Laboratory, University of Paraná, in Curitiba (Brazil), a reference for these studies.

The Jirau model has dimensions of 40 m × 5 m, covering 30 km of the river, being 20 km upstream of the dam and 10 km downstream. It was built with the objective of analyzing sediment transport, as well as the movement of logs and floating debris.

Figure 10.3 Estreito HPP. Mobile bed three-dimensional model. Test results for $Q = 28.700$ m³/s.
Source: (Fudimori, 2013)

Figure 10.4 Water level and velocities measurements.

Figure 10.5 Spillway model. Observe tip scale (red) in water level measurement station.

Figure 10.6 Itaipu HPP model. Institutes LACTEC. Curitiba, Brazil.

Figure 10.7 Itaipu HPP. Three-dimensional model. Paraguay. In the foreground, tip scale (red) for water level measurements in the right bank. In the background, the spillway and the powerhouse.

Figure 10.8 Spillway model. Institutes LACTEC. Curitiba, Brazil. Observe side scale to measure the jet distance.

Figure 10.9 Anta HPP model. Paraíba do Sul river. Rio de Janeiro. Brazil.
Source: (FURNAS Centrais Elétricas S.A.).

Figure 10.10 Jirau HPP model. Sogreah Laboratory. Grenoble, France.

10.5   Hydraulic models of the Tucuruí spillway

Tucuruí HPP, 8.370 MW, was built in a region of metasediments strongly compartmentalized by at least four faults systems, in addition to rock bedding and fracturing plans. The appearance of the rocky massif at the site can be seen in Figure 10.21. The rock is sound at the region of the structures including the plunge pool, except for some discontinuities where the alteration reaches great depths.

The layout of the structures was studied in two hydraulic models at HIDROESB – Hydrotechnical Laboratory Saturnino de Brito (Rio de Janeiro, Brazil):

- one tri-dimensional (Figure 10.11) to study the layout of the works, scale 1:150;
- one bidimensional to study the spillway section, scale 1:50.

Figure 10.11A Three-dimensional reduced model of Tucuruí HPP.

Figure 10.11B Three-dimensional model of Tucuruí spillway. Note jet throw distance
Source: (HIDROESB, Final Report, July 1982).

The studies of the first phase of the plant was carried out between the years of 1976 and 1984, and the studies for the second phase between 1998 and 2001.

In these models were carried out extensive campaigns of tests to test and to optimize:

*   the layout and the positioning of the axis of the dam structures and the approach flow;
*   the structure of the spillway and the structure of the intake/powerhouse
*   the forms of the extreme walls between these structures;
*   the upstream hydraulic spigot over the slope of the central dam at the right side of the spillway
*   (see Figure 10.18);
*   the waterline profiles on the structure;
*   the river diversion in detail.

For the spillway the following tests were made:

*   verification of the conditions of approach of the flow to the structure, with special attention to the extreme spans: on left side is the intake; on the right side there is the rockfill dam;
*   tests of the shape of the structure, crest, Creager profile, pillar shapes and end walls, and the interfaces with the intake and the dam of the central channel, in order to obtain forms that minimize interferences and disturbances/discontinuities on the flow;
*   determination of the waterline profile on the structure for various discharges; these profiles can be estimated using the data in Figure 4.15, when models are not available;
*   verification of the flow capacity of the spillway and of the 40 culverts (see Figures 4.18, 4.20 and 4.21); it should be noted that the spillway discharge coefficient can be estimated using the graphs of Figures 4.10 to 4.14 when no model studies are available;
*   measurements of average and instantaneous pressures on the profile (Figure 4.23); when no model is available the pressures can be estimated as set out in Chapter 4;

Figure 10.12A Tucurruí HPP. Bidimensional model to studies of the spillway. Note the flow lines to $Q = 100.000$ m³/s, without discontinuities.

Source: (Hidroesb, Final Report. 1982).

Figure 10.12B Tucuruí HPP. Bidimensional model to studies of the spillway. Observe jet throw distance, 60 to 70 m, for $Q = 100.000$ m³/s.

Source: (Hidroesb, Final Report. 1982).

Figure 10.13 Tucuruí HPP. Spillway running in prototype. Observe the jet throw distance.

Figure 10.14A Spillway. Upstream view: Flow through the 40 culverts (6,5 m × 13 m).

Figure 10.14B Side view. Note bucket of the spillway. Flow through the 40 culverts (6,5 m × 13 m).

Figure 10.14C Three-dimensional hydraulic model. Downstream view. River diverted through the culverts.

Source: (Hidroesb, Final Report, 1982).

- verification of jet throw distance; when no model studies are available the jet range can be estimated as set out in Chapter 5;
- verification of erosion trends in the area of the plunge pool downstream of the spillway (see Figures 10.20 and 10.21), in mobile bottom, for several discharges spilled;
- detailing of the operation plan of the 23 spillway gates, to be followed by the plant operators.

In addition to these tests, it was possible to test, in detail, the 3rd phase of river diversion by the 40 culverts under the spillway.

Approach of the flow to spillway – abutments

It is important to emphasize the importance of the tests to check the approach to the structure, aiming to obtain a flow with the greatest possible tranquility, that fits and conforms to the concrete structures of the chute, the pillars and the walls, without detachment and depressions of the flow line spilled. This is particularly important in the case of spillway extreme spans where the approach conditions are asymmetrical like in Tucuruí HPP (Tocantins river, Amazon Region, Brazil). In this case we have (as shown in the followings illustrations):

- on the left side of the spillway (V) is the intake (TA): to separate these structures and to better accommodate the flow the VTA wall was provided, showed in Figure 10.15;

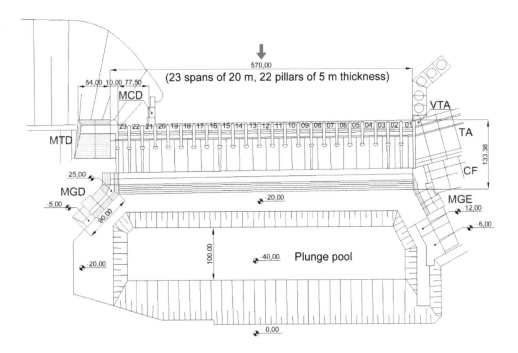

Figure 10.15 Tucuruí HPP. Spillway – plant.

Source: Magela *et al.* (1983).

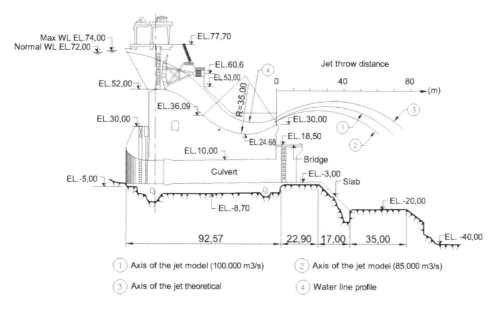

Figure 10.16 Tucuruí HPP. Spillway – cross-section.

Figure 10.17 Tucuruí HPP. Side view. Note VTA ogee, the wall between the intake and the spillway. Just after the spillway, the EHM can be seen. The hydraulic spigot, which is shown in detail in Figure 10.18.

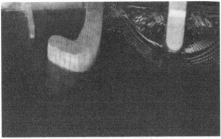
VTA wall – detail in the model.

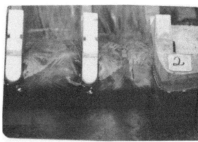
VMTD wall – detail in the model.

EHM – Hydraulic spigot upstream the prototype.

Effect of the EHM on the approach flow in the prototype.

EHM – Detail of the approach flow in the model.

Figure 10.18 Tucuruí HPP: Dam and the spillway in the model and in the prototype.

• on the right side there is the rockfill dam showed in Figure 10.18, where the EHM – upstream hydraulic spigot was foreseen over the slope of the rockfill dam in the central channel of the river.

The operating conditions of these spans were completely different before these structures (VTA and EHM). With the introduction of these special shapes, the flow almost perfectly accommodated to the extreme spans.

The search for the shapes and length of the extreme walls is very laborious. The ogees of these walls usually have quite different shapes, which aim to minimize turbulence in the

Flow details extreme spans in the model.
Close to VMTD wall.

Flow details extreme spans in the model.
Close to VTA wall.

Figure 10.19 Tucuruí HPP: Spillway – flow details close to the extreme walls in the model.

approach flow that can reduce the flow capacity and generate instability in the pressure field (Figure 10.19).

The quieter the approach flow to the spillway, the higher the discharge coefficient and the lower the instabilities on the hydrodynamics pressures on the spill profile, which will minimize the possibility of cavitation occurrence.

The analysis of all these aspects can only be done in hydraulic models which, due to the optimization of the dimensions of the concrete structures, often provide savings for the cost of the project which pay the cost of hydraulic studies in reduced model.

Another important theme studied in the models is the evaluation of the erosive processes downstream of the structures. The whole area susceptible to the scour process is represented in two ways: either with loose moving bottom, or with a certain degree of cohesion. This technique only leads to qualitative estimates of the evolution tendency of the erosion pit.

In the studies with loose moving bottoms, without cohesion, in the modeling of the area that may be subject to the erosive process, crushed rock is used to represent the potentially removable blocks of the rock mass (of course, in the model scale). This type of test applies when, in the region in focus, the mass is very fractured or of low resistance to dynamic efforts. In these tests a good indication of the limit erosion crater is obtained.

In the tests with cohesive moving bottoms, simulation of the resistance of the rock is attempted using, for example, aluminous cement or gypsum. The definition of the mortar

Figure 10.20 Tucuruí HPP spillway. Three-dimensional reduced model. Erosion study in loose moving bottom. Pre-excavation at quota −40,00 m.
Discharges: 35.000, 50.000, 100.000, 85.000 and 50.000 m³/s.

Source: (Hidroesb, Final Report. 1982).

Figure 10.21 Tucuruí HPP spillway. Plunge pool elevation − 40,00 m final project.

trace is complex. We can cite the cases of Grand Rapids (Feldman, 1970) and Picote dam (Lencastre, 1982), where, after lifting the erosion in the prototype, the moving background material of the model was progressively adjusted to reproduce the same degree of change of the bed observed in the work.

The knowledge of real cases of erosion in prototypes, their association with the operating conditions that caused them and the detailed geological survey of the area susceptible to the erosive process, constitute valuable information to better guide the studies of this subject in reduced hydraulic models. It should be recalled that:

- In Chapter 7, Current Efforts Downstream of Energy Dissipators were presented; the cases of the spillways of the: Jaguara HPP (Minas Gerais State, Brazil), where the rock is a quartzite sound, very fractured; and Canoas I HPP (São Paulo/Paraná States, Brazil), where the rock is a dense basalt sound;
- In Chapter 8, Erosion Pit Dimension Assessment Equations were presented; these equations must be used when a reduced model is not available.

Chapter 11

Specific constructive aspects of hydraulic surfaces

With regard to the constructive aspects, this book was only attended with the care taken with the finishes of the hydraulic surfaces submitted to the flow. For aspects related to other constructive aspects, such as the use of sliding forms, users should consult specialized publications.

The hydraulic surfaces of the spillways are subjected to high velocity flows, frequently greater than 30 m/s, and have strict constructive specifications for protection against erosion by cavitation (see section 8.3). Even obeying these specifications accidents has happened, which reinforces the need for enforcement control. For this, every work has its Quality Control Division.

11.1 Tucuruí spillway

For the Tucuruí HPP spillway the construction methods were defined as a function of slope, as shown in the Table 1.1 and in Figure 11.1.

The screeding is done by means of transverse movements of a metallic ruler (tubular). The operation is repeated as many times as necessary. Then the final battening is realized with a wooden trowel (Figures 11.1 and 11.7).

In the 1980s, the pre-assembly, sliding and reinforcement processes were used extensively in the works of the Tucuruí and Xingó plants (Andriolo and Ribeiro, 2015). These developments have spread since the 1990s with the privatization of part of the electricity sector.

Changes in construction technologies have significantly reduced construction times. One can emphasize the intensification of the use of sliding forms, even in massive structures, pre-frame and precast. Likewise, the use of CCR contributed.

The following (Figures 11.2 to 11.9) are some photos that show constructive aspects of spillways. The photographs of the Tucuruí spillway (1980/1981) were extracted from a report by the Quality Division of the Work (copy of the author).

Frame 11.1 Constructive method.

Declivity	Constructive method
Until 1 V:4 H	Screeding
1 V:4 H to 2 V:3 H	Sliding form
Over of 2 V:3 H	Temporary fixed form

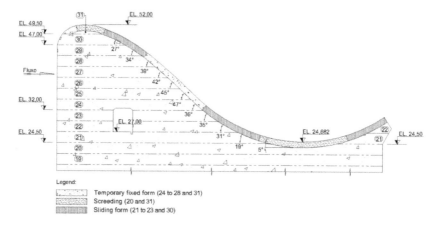

Legend:

Temporary fixed form (24 to 28 and 31)
Screeding (20 and 31)
Sliding form (21 to 23 and 30)

Figure 11.1 Constructive method.

Figure 11.2 Tucuruí HPP – Start of construction of the spillway.
Source: (Eletronorte).

Figure 11.3 Tucuruí HPP – Construction of the spillway. Preparation for concreting the layers 20 and 21 downstream of the bucket, near the takeoff lip.
Source: (Personal archive).

Figure 11.4 Tucuruí HPP – Construction of the spillway. Concreting the layer 21 using sliding form.

Source: (Personal archive).

Figure 11.5 Tucuruí HPP – Construction of the spillway. Correct vibration and placement of the guides (pulleys) to keep the cable parallel to the surface.

Source: (Personal archive).

Figure 11.6 Tucuruí HPP – Construction of the spillway. Form temporarily fixed (layers 24 to 28). Panel removal. Notice surface appearance of concrete.

Source: (Personal archive).

Figure 11.7 Tucuruí HPP – Construction of the spillway. Panel removal. Beginning of the surface finish with wooden trowel.

Source: (Personal archive).

Figure 11.8 Tucuruí HPP – Construction of the spillway.
Source: (CBDB, 2002).

Figure 11.9 Tucuruí HPP – Construction of the spillway.
Source: (CBDB, 2002).

11.2 Jaguara spillway

Jaguara HPP, 424 MW, Grande river, Minas Gerais/São Paulo, Brazil.

Figure 11.10 Jaguara HPP. Spillway. Downstream channel refurbished.
Source: (CBDB, 2002).

Figure 11.11 Jaguara HPP. Remodeling of the channel downstream of the spillway. April to September 2001.
Source: (CBDB, 2002).

Figure 11.12 Jaguara HPP. Construction of the channel downstream of the spillway.
Source: (CBDB, 2002).

11.3 Mascarenhas de Moraes spillway

Mascarenhas de Moraes (Peixoto) HPP, 476 MW, Grande river, Minas Gerais/São Paulo, Brazil.

Figure 11.13 Mascarenhas de Moraes HPP. Construction of the complementary spillway, 1998/
2002. See Figures 1.10 to 1.11.

Source: (CBDB, 2002).

11.4 Heart Butte spillway

The following illustrations shows the dam and aspects of the construction of the conduit and stilling basin for a combined drop inlet of the morning glory spillway of the Heart Butte dam on the Heart river in North Dakota (USA). Length of the dam = 564 m; H = 43 m; Q = 161 m^3/s.

Figure 11.14 Heart Butte dam. Construction of the morning glory spillway. See also Figures 2.2
and 2.8 b.

Source: (DSD, 1974, Figure 241).

11.5   El Guapo spillway

The following illustrations present aspects of the reconstruction of the El Guapo dam spillway in Venezuela, which ruptured due to insufficient flow capacity. The dam is 60 m high and 524 m long, with a spillway to 102 m³/s (see section 1.6).

Figure 11.15 Reconstruction of the El Guapo spillway (Venezuela).

Figure 11.16 Reconstruction of the El Guapo dam spillway (Venezuela).

11.6   Belo Monte spillway

Belo Monte HPP 11.000 MW, Xingu river (Pará State, Brazil). Spillway: $Q = 62.000$ m³/s; 20 gates 20 m × 22,30 m.

Figure 11.17 Belo Monte HPP. Dam, spillway and auxiliary powerhouse

11.7 Colíder spillway

Colíder HPP, 300 MW, Teles Pires river (Mato Grosso state, Brazil). Capacity of the spillway: 6.673 m³/s. 4 gates 12 m × 19 m. (COPEL).

Figure 11.18 (A) Colíder HPP, 300 MW. (B) Spillway, downstream view. (C) Gate detail.

11.8   São Manoel spillway

São Manoel HPP, 700 MW, Teles Pires river (Mato Grosso state, Brazil). H = 22,68 m. Spillway: 3 gates and capacity of 13.828 m³/s (EESM-EDP Brasil, FURNAS, CTG).

(A)

(B)

Figure 11.19 São Manuel HPP. (A) River diversion. (B) Power plant and spillway.

11.9   Santo Antônio spillway

Santo Antônio HPP, 3.568 MW, 50 bulb turbines, H = 13,9 m, Madeira river (Rondônia, Brazil). Capacity of the spillways: 84.000 m³/s. 15 gates 20 m × 24 m (Santo Antônio Energia).

Figure 11.20 Santo Antônio HPP. Downstream view of the works. From left to the right: trunk interception system; powerhouse; spillway 2; fish ladder; powerhouse; main spillway; powerhouse (left bank).

Figure 11.21 Santo Antônio HPP. Historic flood (2014).

11.10 Mauá spillway

Mauá HPP, Tibagi river (Paraná state, Brazil). 361 MW; H = 85 m; L =775 m; Spillway: 9.368 m³/s; 4 gates 11,5 m × 18,0 m. (Consórcio Energético Cruzeiro do Sul – CECS).

morozcomunicação.com.br

www.youtube.com

www.portal.ortigueira.pr.gov.br

Figure 11.22 Mauá HPP, Tibagi river (Paraná state, Brazil).

References

Abbas, A. & Zeinab, T. (2012) Fusegates selection and operation simulation-optimization approach. *IWA Publishing, Journal of Hydroinformatics*, 14(2).

Abecasis, F. M. M. (1961) Spillways: Some Special Problems. LNEC: National Laboratory of Civil Engineering. Memory 175. Lisbon.

Afshar, A. & Marino, M. A. (1990) Determining optimum spillway capacity based on estimated flood distribution. *WP&DC*, January.

Akmedov, T. (1968) Local erosion of fissured Rock at the downstream end of a spillways. *Hydrotechnical Construction*, 9, 821–824.

Aksoy, S. & Ethembabaoglu, S. (1979) Cavitation Damage at the Discharge Channels of Keban Dam. Q.50, R. 21, 13° ICOLD. New Delhi.

Alla, A. A. (1996) The role of fusegates in dam safety. *Hydropower & Dams, Wallington*, 6.

Allen, G. C. (2010) USACE – U. S. Army Corps of Engineers: Sayano-Sushenskaya. *Southwestern Federal Hydropower Conference, Branson, Missouri.*

Amorim, J. C. C. (2012) Module 1: Dams. Legal, Technical and Socio-Environmental Aspects, Unit 7: Hydromechanical Aspects. Course on Dams Safety. Org.: Itaipu Binational, ANA: National Water Agency and MME: Ministry of Mines and Energy, Brazil.

Anami, K., et al. (2013) Dynamic stabilization of Folsom dam tainter-gate by replacing hoist with cables. *11th International Conference on Vibration Problems, 9–12 September, Lisbon.*

Anderson, D., Paton, T. A. L. & Blackburn, C. L. (1960) Zambesi hydro-electric development at Kariba, first stage. *Proc. ICE – Inst. of Civil Engineers*, 17(1), 1 September. London.

Andriolo, F. R., & Ribeiro, L. E. F. (2015) Aspects of Constructive Methodologies Adopted in Dam Works under the Optics of Programmatic Quality-Advantages-Care. 30th SNGB. Foz do Iguaçu, Paraná State, Brazil.

Atkinson, C. H. & Overbeeke, K. (1973) Design and Operation of Lower Notch Spillway. Q.41, R.53, 11° ICOLD, Madrid.

Atkinson, C. H. & Overbeeke, K., 11th ICOLD (1973) Control of Flows and Energy Dissipation during Construction and after Construction, Madrid.

Aubin, L., et al. (1979) Les Evacuateurs de Crue des Amenagements Hydroelectriques LG2 et LG1 du Complexe La Grande. Q.50, R.8, 13° ICOLD, New Delhi.

Ball, J. W. (1963) Construction Finishes and High-Velocity Flow. ASCE, JCD, CO2, September.

Ball, J. W. (1976) Cavitation from Surface Irregularities in High-Velocity Flow. ASCE, JHD, HY9, September.

Balloffet, A., Gotelli, L. M. & Meoli, G. A. (1948) Hydraulics: Civil Engineers of the University of Buenos Aires. Ediar Soc. Anón. Editores, Buenos Aires.

Berryhill, R. H. (1957) Stilling basin experiences of the corps of engineers. *Journal of the Hydraulics Division, ASCE*, 83, HY 3, June.

Berryhill, R. H. (1963) Experience with prototype energy dissipators. *Journal of the Hydraulics Division, ASCE*, 89, HY 3, May.

Binger, W. V. (1972) Tarbela dam project, Pakistan: Journal of the Power Division. *Proceedings of the ASCE, October, New York.*

Bobko, G. (1980) Particularities of the Flow in Compact Stilling Basins. 13th SNGB. Rio de Janeiro, Brazil.

Bodhaine, G. L. (1968) Measurement of Peak Discharge at Culverts by Indirect Methods. Techniques of Water-Resources Investigations of the United States Geological Survey. Book 3, Applications of Hydraulics.

Bollaert, E. F. R. (2004a) A comprehensive model to evaluate scour formation in plunge pools. *International Journal of Hydropower & Dams*, (1), 94–101.

Bollaert, E. F. R. (2004b) A new procedure to evaluate dynamics uplift of concrete linings of rock blocks in plunge pools. *Proceedings of the International. Conference on Hydraulics of Dams and River Structures*, Yazdandoost & Attari (eds.), Teheran, pp. 125–132.

Bollaert, E. F. R. (2006) The comprehensive scour model: Theory and feedback from practice. *Aqua-Vision.* www.aquavision-eng.ch

Bollaert, E. F. R. (2010) A prototype-scaled rock scour prediction model. *30th International Conference on Scour and Erosion (ICSE), USSD: United States Society on Dams, April, Sacramento.*

Bollaert, E. F. R. & Petry, B. (2006) Application of a physic based scour prediction model to Tucuruí dam spillway (Brazil). *Proc. of the Third Intern. Conference on Scour and Erosion, Amsterdam.*

Bollaert, E. F. R., et al. (2012) Kariba dam plunge pool scour: Quase-3D numerical predictions. *International Conference on Scour and Erosion 6 (ICSE), 27–31 August, Paris.*

Bornschein, A., POHL, R. Dam Break During the Flood in Saxony/Germany in August 2002. *Proc. IAHR Congress. Thessaloniki, Greece*, 2003.

Bornschein, A. & Pohl, R. (2003) Dam break during the flood in Saxony/Germany in August 2002. *Proc. IAHR Congress, Thessaloniki, Greece.*

Bowles, D. S. (2007) Tolerable risk for dams: How safe is safe enough. *US Society on Dams Annual Conference, March*, Philadelphia, Pennsylvania.

Brater, E. F. & King, H. W. (1976) *Handbook of Hydraulics, for the Solution of Hydraulic Engineering Problems*, 6th edition. McGraw-Hill Book, University of Michigan, USA.

Brito, S. N. (1991) Geomechanical Investigations Downstream of Spillways with Skijump. 19th SNGB. Aracaju, Brazil.

Carter, R. W. (1957) Computation of Peak Discharge at Culverts. US Geological Survey Cir.376 (57).

CBDB (1978) Topmost Dams of Brazil, Heavy Construction Magazine, Rio de Janeiro, Brazil.

CBDB (1982a) Main Brazilian Dams, Design, Construction and Performance, Rio de Janeiro, Brazil .

CBDB (1982b) Proceedings of the International Symposium on General Arrangements of Dams in Narrow Gorges. CBDB, Rio de Janeiro, Brazil.

CBDB (2000) Main Brazilian Dams, Design, Construction and Performance, Volume II, Rio de Janeiro, Brazil.

CBDB (2002) Large Brazilian Spillways: An Overview of Brazilian Practice and Experience in Designing and Building Spillways for Large Dams. Rio de Janeiro, Brazil.

CBDB (2006) Highlights of Brazilian Dam Engineering, Rio de Janeiro, Brazil.

CBDB (2009) Main Brazilian Dams, Design, Construction and Performance, Rio de Janeiro, Brazil.

CBDB (2011) The History of Dams in Brazil. 19th, 20th and 21st centuries. Fifty Years of the Brazilian Dam Committee, Rio de Janeiro, Brazil.

Cedergren, H. R. (1967) *Seepage, Drainage and Flownets*. John Wiley & Sons, Inc., Hoboken, NJ.

CFE, Comissión Federal de Electricidad (2014) Revisón de las Presas a Cargo de la CFE, 4 December.

Chanson, H. (1993) Stepped spillway flows and air entrainment. *Canadian Journal of Civil Engineering*, 117(3), 349–372.

Chanson, H. (1994) Hydraulics of skimming flows over stepped channels and spillways. *Journal of Hydraulic Division, IAHR*, 32(3), 445–460.

Chanson, H. (1995) *Hydraulic Design of Stepped Cascades, Channels, Weirs and Spillways*. Pergamon, an Imprint of Elsevier Science, Oxford, UK.

Chanson, H. (2001) Hydraulic design of stepped spillways and downstream energy dissipators. *Dam Engineering*, 11(4).

Chanson, H. (2004) *The Hydraulic of Open Channel Flow: An Introduction*, 2nd edition. Butterworth-Heinemann, Elsevier, Burlington, MA, USA.

Chanson, H. (2009) *Embankment Overflow Protection Systems and Earth Dam Spillways: Chapter 4 of Dams: Impacts, Stability and Design*. Nova Science Publishers, Inc., United Kingdom.

Chavarri, G., et al. (1979) Spillway and Tailrace Design for Raising of Guri Dam Using Large Scale Hydraulic Model. 13° ICOLD, Q.50, R.12, New Delhi.

Chow, V. T. (1959) *Open Channel Hydraulics*. McGraw-Hill Book, New York.

Christodoulou, G. C. (1993) Energy dissipation on stepped spillways. *Journal of Hydraulic Engineering, ASCE*, 119(5), May.

CIRIA – Construction Industry Research and Information Association (1978) The Hydraulic Design of Stepped Spillways, Report n. 33, London.

Cola, R. (1965) Energy Dissipation of a High Velocity Vertical Jet Entering a Basin. IAHR, 11°Congress, Leningrad.

Creager, W. P., Justin, J. D. & Hinds, J. (1945) *Engineering for Dams*. John Wiley & Sons, New York.

Cunha, L. V., Lencastre, A., et al. (1966) La Dissipation de L'Energie Dans um Évacuateur em Saut de Ski. Observation de L'Érosion. LNEC, Memory No. 288, Lisboa.

Dai Qing (1989) The River Dragon Has Come. YI SI, Chapter Three.

Dao-Yang, D. & Man-Ling, L. (1979) Mathematical Model of Flow Over a Spillway Dam. 13° ICOLD, Q.50, R.55, New Delhi.

Davis, C. V. (1952) *Handbook of Applied Hydraulics*. McGraw-Hill Book, New York.

Demiroz, E. & Ivazoglu, M. (1991) The Remedial Works for the Spillway Discharge Channels of Keban Dam. 17° ICOLD, Q.65, R.63, Vienne.

Design of Small Dams (1974)

Eletrobras (1987) *Guide to Calculate the Flood of Spillway Design*. Eletrobras, Holding Company of the Power Sector, Rio de Janeiro, Brazil.

Eletrobras/CBDB (2003) *Criteria of Civil Project of Hydroelectric Power Plants*. Rio de Janeiro, Brazil. www.eletrobras.com

Elevatorski, E. A. (1959) *Hydraulic Energy Dissipators*. McGraw-Hill Book, USBR, Denver, USA.

Erbisti, P. C. F. (2004) *Design of Hydraulic Gates*. A. A. Balkema Publishers, Netherlands.

Escande, L. (1929) *Étude théorique et expérimentale sur la similitude des fluides incompressible pesants*. Toulouse.

Falvey, H. T. (1990) Air-Water Flow in Hydraulic Structures. Engineering Monograph 41, USBR.

Falvey, H. T. (1990) Cavitation in Chutes and Spillways. Engineering Monograph 42, USBR.

Feldman, G. M. (1970) Stabilization of Spillway Scour Hole at Grand Rapids, Manitoba. Hydraulic Power Section, Canadian Electrical Association, International Inn, Winnipeg, Manitoba, 24 September.

FEMA, Federal Emergency Management Agency, US Department of Homeland Security (2014) Technical Manual: Overtopping Protection for Dams. FEM P-1015, May.

Fudimori, M. H., et al. (2013) Structural Measures in Dissipation Basins to Reduce Erosion Downstream of the Energy Dissipator. 29th SNGB, Porto de Galinhas, Brazil.

Furstenburg, L., et al. (1991) The Influence of Foundation Conditions on Spillway and Plunge Pool Design at Katske Dam. 17° ICOLD, Q.66, R.93, Vienne.

Gaëtan, H. (2011) Summary of the Knowledge Acquired in Northern Environments from 1970 to 2000. Hydro-Québec, September, Montréal, Canada.

Galindez, A., et al. (1967) Spillways in a Peak Flow River (Douro and its tributaries). 9° ICOLD, Q.33, R.22, Istamboul.

Ginocchio, R. et al. (2012) L'Énergie Hydraulique. Collection EDF R&D. Lavoisier.

Grishin, M. M. (1982) *Hydraulic Structures*, Volumes 1 & 2. Mir Publishers, Moscow.

Guinea, P. M., et al. (1973) Selection of Spillways and Energy Dissipators. 11° ICOLD, Q.41, R.66, Madrid.

Hampton, I., Amghar, M. & Wiley, J. (2004) Hydraulic model study for Eildon dam improvement project. *ANCOLD/NZSOLD Conference*, Brisbane, Australia.

Hartung, F. (1973) Gates in Spillways of Large Dams. 11° ICOLD, Q.41, R.72, Madrid.

Hartung, F. & Häusler, E. (1973) Scours, Stilling Basins and Downstream Protection Under Free Fall Overfall Jets at Dams. 11° ICOLD, Q.41, R.3, Madrid.

Häusler, E. (1983) *Spillways with High Energy Concentration: International Symposium on General Arrangement of Dams in Narrow Gorges*. CBGB, Rio de Janeiro.

Hay, N. & Taylor, G. (1970) *Performance and Design of Labyrinth Weir*. JHD-ASCE, NYC, USA.

Hayeur, G. (2001) *Summary of the Knowledge Acquired in Northern Environments from 1970 to 2000*. Hydro-Québec, Montréal, September. Available on the internet version .pdf.

HDC – Hydraulic Design Criteria: Corps of Engineers, Waterways Experiment Station, U. S. Army Engineer (1955/1959) Vicksburg, Mississippi. Available on the internet version .pdf.

HDS – Hydraulic Design of Spillways: Engineer Manual EM 1110-2-1603, Headquarters, Department of the Army, Office of the Chief of Engineers (1965). Available .pdf file on the internet.

HDSBED – Hydraulic Design of Stilling Basin and Energy Dissipators (1983) Engineering Monograph 25, USBR. Available on the internet version .pdf.

Hidraulica. Balloffet, A. et al. (1952) Ediar Soc, Anón. Editores. Buenos Aires. Argentina.

Hinds, J. (1926) Side channel spillway: Hydraulic theory, economic factors and experimental determination of losses. *Transactions, ASCE*, 89, 881–939.

Hinks, J. L. & Charles, J. A. (2004) *Reservoir Management, Risk and Safety Considerations*. British Dams Society, UK.

Hydraulic Design Criteria. (1955) Corps of Engineers, USA.

Hydraulic Design. Lysne, D. et al. (2003) Norwegian University of Science and Technology.

ICOLD (1987) Spillways for Dams. Bulletin No 58.

ICOLD (1995) Dam Failures, Statistical Analysis, Bulletin 99.

ICOLD (2012) Safe Passage of Extreme Flood, Bulletin 142.

IDROTECNICA, Organo Ufficiale Dell'Associazione Idrotecnica Italiana (1988) Special Issue for the 16° International Congress on Large Dams, San Francisco, Maggioli Editore, March–April.

Jabara, M. A. and Legas, J. (1986) Selection of Spillways, Plunge Pools and Stilling Basins for Earth and Concrete Dams. 11° ICOLD, Q.41, R.17, Madrid.

Jansen, R. B. (1983) *Dams and Public Safety: A Water Resources Technical Publication*. US Department of the Interior, Bureau of Reclamation, USBR, Denver, USA.

Jansen, R. B. (1988) *Advanced Dam Engineering for Design, Construction and Rehabilitation*, USBR, Denver, USA.

Johnson, R. B. (1973) Spillway Types in Australia and Factors Affecting Their Choice. 11° ICOLD, Q.41, R.46, Madrid.

Junshou, M. (1982) *The Layout and Energy Dissipation for Spillways of Several Hydroelectric Power Projects in Deep Valleys with Large Discharge and High Drop in Head (China): International Symposium on General Arrangement of Dams in Narrow Gorges*. CDBD, Rio de Janeiro.

Khader, M. H. A. & Elango, K. (1974) Turbulent pressure field beneath a hydraulic jump. *Journal of Hydraulics Research*, 2(4), Set.

Kohler, W. H. (1969) Selection of outlet works gates and valves. *ASCE Annual and Environmental Meeting, Chicago, IL*. Available from US Bureau of Reclamation.

Kollgaard, E. B. & Chadwick, W. L. (1988) Development of Dam Engineering in the Unites States. Prepared in Commemoration of the 16° ICOLD, 1988, San Francisco (USA).

Kraichnan, R. H. (1956) Pressure fluctuations in turbulent flow over a flat plate. *The Journal of the Acoustical Society of America*, 28(3), May.

Kramer, K. (2004) *Development of Aerated Chute Flow: A Thesis Submitted to the Swiss Federal Institute of Technology Zurich for the Degree of Doctor of Technical Science*, Swiss Federal Institute of Technology – Zurich.

Lee, H., Koh, D. & Yum, K. (2008) Enhancement program for hydrologic safety of existing dams in Korea. *28th Annual USSD Conference, 28 April–2 May, Portland, Oregon*.

Lees, P. & Thomson, D. (1997) Emergency management, Opuha dam collapse, waitangi day. *IPENZ Proceedings of Technical Groups 30/2 (L/D), NZSOLD – New Zealand Society of Large Dams Conference, Wellington.*

Lemos, F. O. (1965) *The Instability of the Boundary Layer: Its Influence on the Design of the Dam Dischargers.* PhD Final Thesis, LNEC, Lisbon.

Lemos, F. O. (1983) Behavior of the structures of some dams built on narrow vals. *International Symposium on General Arrangement of Dams in Narrow Gorges, Rio de Janeiro (1982) and LNEC, Memory 580, Lisbon.*

Lemos, F. O. (1986) Criteria for the Hydraulic Design of the Zone of Approach of the Flow to the Structures of Flood Discharges. XII Latin American Congress of Hydraulics, São Paulo, Brazil.

Lencastre, A. (1972) *General Hydraulics: Edgard Blücher Ltda. I.S.T.* Published with the Collaboration of University of São Paulo, Lisbon.

Lencastre, A. (1982) Spillways with high concentration of energy. *CBDB/ICOLD, Report of the Theme in the Proceedings of the International Symposium on General Arrangements of Dams in Narrow Gorges, CBDB, Rio de Janeiro, Brazil.*

Leyland, B. (2008) A new system for raising spillway gates. *Hydropower & Dams,* (3).

Leyland, B. (2010) Lessons from the accident at the Sayano-Shushenskaya hydro power station. *EEA Conference & Exhibition, Christchurch.*

Leyland, B. (2014a) *Gated Spillways: Are They Safety Enough.* The British Dam Society, London.

Leyland, B. (2014b) The safety of spillway gates. *Hydropower & Dams, Conf. Lake Como, Italy.*

Liggett, J. A. (1994) *Fluid Mechanics.* McGraw-Hill, Inc., USA.

Linsley, R. K., Jr. & Franzini, J. B. (1955) *Elements of Hydraulic Engineering.* McGraw-Hill, Inc., USA.

Lopardo, R. A. (1987) Notes on Macro-Disturbance Pressure Fluctuations, Analysis and Application to Hydraulic Jump. Latin American Hydraulics Magazine, n. 2, São Paulo.

Lopardo, R. A. & Sollari, H. G. (1980) Pressure Fluctuation in Hydraulic Jump. Latin American Hydraulics Congress, 9, IAHR, Mérida.

Lowe, J., III, et al. (1979) Tarbela Service Spillway Plunge Pool Development. WP&DC, November.

Lux, F., III & Hinchliff, D. L. (1985) Design and Construction of Labirinth Spillways 15° ICOLD, Q.59, R.15, Lausanne.

Lysne, D. K., et al. (2003) *Hydropower Development, Hydraulic Design,* Volume 8. NUST – Norwegian University of Science and Technology, Department of Hydraulic and Environmental Eng, Trondheim.

Magela, G. P. (2003) Criteria to the Design of Hydraulic Structures Aiming to Avoid Cavitation Erosion. 25th SNGB, Salvador, Brazil.

Magela, G. P. (2015a) *Design of Hydroelectric Power Plants Step by Step (Step by Step Hydroelectric Power Plants Project).* Oficina de Textos, São Paulo, Brazil.

Magela, G. P. (2015b) Information on Theodore Roosevelt (USA) and Kariba (Africa) Dams Reform for Security Operation. 30th SNGB, Foz do Iguaçu, Brazil.

Magela, G. P. & Brito, S. N. (1991) The Use of a Skijump Spillway in Rocky Massifs Are Very Fractured: The Case of Jaguara HPP. 19th SNGB. Aracajú, Brazil.

Magela, G. P. & Brito, S. N. (1996) Erosion in Stilling Basins: Hydraulic and Geotechnical Aspects. CBGB Magazine, Publication n. 03/96, Rio de Janeiro, Brazil.

Magela, G. P. & Rodrigues, J. L. P. (1991) Erosions Downstream of Hydraulic Structures: Performance of Some Spillways of CESP. 19th SNGB, Aracajú, Brazil.

Magela, G. P., et al. (1983) Spillway of Tucuruí HPP. *Luso-Brazilian Symposium on Simulation of Hydraulic Modeling and Water Resources, Blumenau, Brazil.*

Magela, G. P., Lopes, M. L. & Araújo, A. L. (2015) Thirty Years of Operation of the Tucuruí HPP Spillway. 30th SNGB, Foz do Iguaçu, Brazil.

Marques, M. G. (1995) Nouvelle Approche Pour le Dimensionnement des Dissipateurs à Auge. Doctoral Thesis, Laval Univesity, Québec, Canada.

Marques, M. G., et al. (1995) Pressure fluctuation in hydraulic jump. *In Latin American Hydraulics Congress, 17, IAHR, Guayaquil.*

Martins, R. (1973) Erosive Action of Free Jets Downstream of Hydraulic Structures. LNEC Memory 486, Lisbon.

Martins, R. (1975) Scouring of Rock River Beds by Free Jet Spillways. WP&DC, April.

Martins, R. (1977) Kinematics of the Free Jet in the Scope of Hydraulic Structures. LNEC Memory 424, Lisbon.

Martins, R. (1981) Hydraulics of Overflow Rockfill Dams. LNEC Memory 559, Lisbon.

Martins, R. (1986) Preliminary experimental design and studies of a rock spillway. *XII Congress Latin American Hydraulics, ABRH, São Paulo, Brazil.*

Mason, P. J. (1982) The choice of hydraulic energy dissipator for dam outlet works based on a survey of prototype usage. *Proc. of Institute of Civil Engineers, Part 1, 72, London.*

Mason, P. J. (1984) Erosion of plunge pools downstream of dams due to the action of free-trajectory jets. *Proceedings of Institute of Civil Engineers, Part 1, 76, London.*

Mason, P. J. (1986) Estimating Plunge Pool Scour. Discussion. WP&DC, February.

Mason, P. J. (1993) Practical Guidelines for the Design of Flip Buckets and Plunge Pools. WP&DC, September/October.

Mason, P. J. & Arumugan, K. (1985) Free jet scour below dams and flip buckets. *ASCE, JHE, 3(2),* February.

Mason, P. J. & Arumugan, K. (1986) A review of 20 years of scour development at Kariba dam. *2nd Conference of Flood and Flood Control, Cambridge.*

Mays, L. W. (1999) *Hydraulic Design Handbook.* McGraw-Hill Handbooks, New York.

Micovic, Z. (2014) *Flood Hazard for Dam Safety: Where the Focus Should Be.* Hidrovision, USA.

Miller, D. S. (1978) *Internal Flow Systems.* British Hydromechanics Research Association, United Kingdom.

Minami, I. (1970) A Consideration of the Supervision of a Concrete Arch Dam in Flood Time. 12° ICOLD, Q.48, R.8, Montreal.

Mitchell, W. R. (1973) River Diversion Arrangements for the Cethana Power Scheme. 11° ICOLD, Q.41, R.9, Madrid.

Moraes, H. M. et al. (1983) Vertedouro da UHE Tucuruí. Simpósio Luso-Brasileiro sobre Simulação de Modelação Hidráulica e Recursos Hídricos. Blumenau, Santa Catarina, Brazil.

Moraes, J. (1979) Selection of Basic Design of Itaipu Spillway. 13° ICOLD, Q.50, R.14, New Delhi.

Neidert, S. H. (1980a) CEHPAR: Hydraulics and Hydrology Center Prof. Parigot de Souza. Spillways. Power Dissipation. Cavitation and Erosion. Publication n. 37, Curitiba, Brazil.

Neidert, S. H. (1980b) General Report on the theme "Spillway Performance – Energy Dissipation, Cavitation and Erosion," 13th SNGB, Rio de Janeiro, Brazil, April.

Noret, C., et al. (2012) Kariba Dam on Zambezi River: Stabilizing the Natural Plunge Pool. *6th ICSE – International Conference on Scour and Erosion, 27–31 August, Paris.*

Novak, P., et al. (2007) *Hydraulic Structures.* Taylor and Francis, New York.

Oliveira, A. R. & Leme, C. R. de M. (1985) Adding 1.000m^3/s to Euclides da Cunha Outflow. 15° ICOLD, Q.59, R.7, Lausanne.

Oliveira, A. R., et al. (1985a) Case Histories of Repairs of Concrete Surfaces Subjected to Water Erosion. 15° ICOLD, Q.59, R.5, Lausanne.

Oliveira, A. R., et al. (1985b) Erosion on the Left Wrap-Around of José Ermírio de Moraes Dam (Água Vermelha). 15° ICOLD, Q.59, R.6, Lausanne.

Oliveira, D. D. (1986) Hydraulic Design of a Morning Glory Spillway by the Wagner and Lazzari Methods: Comparative Analysis. *XII Congress Latin American Hydraulics, ABRH, São Paulo, Brazil.*

Oskolov, A. G. & Semenkov, V. M. (1979) Experience in Designing and Maintenance of Spillways Structures on Large Rivers in the USSR. 13° ICOLD, Q.50, R.46, New Delhi.

Peltier, Y., et al. (2015) Pressure and velocity on an ogee spillway crest operating at high head ratio: Experimental measurements and validation. *2nd International Workshop on Hydraulics Structures, 8–9 May, Coimbra.*

Peterka, A. J. (1983) Hydraulic Design of Stilling Basin and Energy Dissipators. EM 25, USBR (83).

Pinto, A. V. & Martins, R. (2011) Barcelos dam: A self-spillway Rockfill dam. Second *International Symposium on Rockfill Dams, Rio de Janeiro.*

Pinto, L. C. S. (1991) Identification in hydrodynamic effort model. *19th SNGB, Aracajú, Brazil.*

Pinto, L. C. S., et al. (1988) Experimental analysis of pressure fluctuation on the basis of a free and drowned hydraulic jump. *Latin American Hydraulics Congress, 13, Havana, IAHR.*

Pinto, N. L. (1979) Cehpar: Hydraulics and Hydrology Center Prof. Parigot de Souza. Cavitation and Aeration in High Speed Flows. Publication n. 35, Curitiba.

Pinto, N. L. (1983) Energy dissipation and erosion downstream of dams. *Symposium on the Geotechnics of the Alto Paraná Basin, ABGE, São Paulo.*

Pinto, N. L. (1989) Designing Aerators for High Velocity Flow. WP & Dam Construction, July.

Pinto, N. L. (1991) *Prototype Aerator Measurements in Air Entrainment in Free-Surface Flows.* Wood, I. R. (ed.). University of Canterbury, New Zealand. A. A. Balkema.

Pinto de Campos, J. A. (1963) Hydrodynamic Action of a Discharge Nappe on Fragments of a Rocky Bed and Conditions of Breakage of This. Yuditskii, G. A. Izvestiya VNII Gidrotekhniki. Moskva, Leningrad, 72, pp. 35–60. LNEC, Translation n. 442. Lisbon (1983).

Pravdivets, Y. P. (1989) Stepped Protection Blocks for Dam Spillways, WP&DC, July.

Q.50, 13th ICOLD (1979) Spillways and Downloads of Large Capacity Fund, New Delhi.

Quintela, A. C. and Cruz, A. A. (1982) Cahora-Bassa Dam Spillway. Conception, Hydraulic Model Studies and Prototype Behavior. *The Transactions of the International Symposium on Layout of Dam in Narrow Gorges.* Rio de Janeiro, Brazil.

Quintela, A. C. & Ramos, C. M. (1980) Protection against Erosion of Cavitation in Hydraulic Works. LNEC. Memory 539, Lisbon.

Ramos, C. A. M. (1978) *Macro Turbulence of Crushed Flows in Energy Dissipation Structures.* Thesis Presented in the LNEC Specialist Contest, Lisbon.

Rao, K. N. S. (1982) Design of energy dissipators for large capacity spillways. *International Symposium on General Arrangement of Dams in Narrow Gorges, CBGB, Rio de Janeiro, Brazil.*

Rehbock, T. (1929) *Transactions, ASCE*, Volume 93. p. 1143, ASCE.

Regan, R. P., et al. (1979) Cavitation and Erosion Damages of Sluices and Stilling Basins at two High-Head Dams. Q. 50, R.11, 13° ICOLD, New Delhi.

Reinius, E. (1970) Head Losses in Unlined Rock Tunnels. WP&DC, England, July–August.

Reinius, E. (1986) Rock Erosion. WP&DC, England, June.

Ribeiro, A. A., et al. (1967) Erosion in Concrete and Rock Due to Spillway Discharges. 9° ICOLD, Q.33, R.19, Istamboul.

Ribeiro, A. A., et al. (1973) Bed Protection Downstream of a Big Dam Founded in Alluvia. 11° ICOLD, Q.41, R.38, Madrid.

Ribeiro, A. A., et al. (1976) Fundamental Considerations on the Planning of Concrete Dams Founded in Alluvia, in Large Flow Rivers. Crestuma Dam. 12° ICOLD, Q.46, R.17, Mexico.

Ribeiro, A. A., et al. (1979) Fundamental Studies on the Planning of Construction in Large Flow Rivers with River Bed in Alluvia. Crestuma Dam. 13° ICOLD, Q.50, R.23, New Delhi.

Rouse, H. (1950) *Engineering Hydraulics.* John Wiley & Sons, Inc., New York.

Rouse, H. (1938/1961) *Fluid Mechanics for Hydraulic Engineers.* Dover Publications, New York.

Rouse, H. (2011) Fluid Mechanics for Hydraulic Engineers. Abdul Press.

Rudavsky, A. B. (1976) Selection of Spillways and Energy Dissipators in Preliminary Planning of Dam Developments. 12° ICOLD, Q.46, R.9, Mexico.

Sanagiotto, D. (2003) *Characteristics of Spill on Stepped Spillways in Declivity Steps 1V: 0,75H.* Masters dissertation, IPH/UFRGS, Porto Alegre, Brazil.

Saville, T., et al. (1962) Freeboard allowances for waves in inland reservoirs. *Journal of Hydraulic Engineering, ASCE*, 88(2). New York (62).

Schoklitsch, A. (1932) *Hydraulic Structures: A Text and Handbook.* Published by ASME – The American Society of Mechanical Engineers, USA.

Schreiber, G. P. (1978) *Usinas Hidrelétricas – Engevix Estudos e Projetos de Engenharia.* Editora Edgard Blücher Ltda, São Paulo.

Scimemi, E. (1937) *Il Profilo Delle Digne Sfiranti*. L'Energia Elettrica, Italy.

Serafini, J. S. (1963) Wall Pressure Fluctuations and Pressure Velocity Correlations in a Turbulent Boundary Layer. NASA R-165.

Sharma, H. R., et al. (1982) Selection of type and location of dam power houses. *International Symposium on General Arrangement of Dams in Narrow Gorges, CBDB, Rio de Janeiro, Brazil.*

Simões, A. (2008) *Considerations on Hydraulics of Spillways in Steps: Non-Dimensional Methodologies for Pre-Dimensioning*. Masters Dissertation, School of Engineering of São Carlos, University of São Paulo, Brazil.

Sobrinho, J. A. & Infanti, Jr., N. (1986) Erosion of rock masses subject to flow actions: Some geomechanical and hydraulic aspects. *V Int. Cong. Engineering Geology, Buenos Aires.*

Sobrinho, J. A., et al. (2000) Dona Francisca hydropower plant: Free overflow spillway with energy dissipation trough steps on RCC dam. *International Workshop – Hydraulics of Stepped Spillways, ETHZ, Zurich.*

Soos, I. G. K. (1982) Proposed lay-out and construction in a Narrow Gorge. *Proceedings of the International Symposium on General Arrangement of Dams in Narrow Gorges, CBDB, Rio de Janeiro.*

Srinivas, N. (2015) Failure of spillway radial gate of Narayanapur dam: A case study. *First National Dam Safety Conference, Technical Session 6*, 25 March.

Spurr, K. J. W. (1985) Energy approach to estimating scour downstream of a large dam. *WP&DC, July, England.*

Stanchev, S., et al. (1973) Flood Problems in Dam Construction and Operation in the Peoples's Republic of Bulgaria. 11° ICOLD, Q.41, R.12, Madrid.

Stephenson, D. (1979) Gabion Energy Dissipators. Q.50, R. 3, 13° ICOLD. New Delhi.

Stutz, R. O., et al. (1979) The Skijump of the Karakaya Hydroelectric Scheme. 13° ICOLD, Q.50, R.33, New Delhi.

Suzuki, Y. (1973) Design of a Chute Spillway Jointly Serving as the Roof Slab of a Hydro Power Station and Its Review on the Vibration During Flood. 11° ICOLD, Q.41, R.21, Madrid.

Szpilman, A., et al. (1991) Itaipu spillway and region downstream of the trampoline: Hydraulic behavior after 8 years of operation, *19th SNGB, Aracajú, Brazil.*

Taraimovich, I. I. (1978) Deformation of channels below high head spillways on Rock Foundations. *Hydrotechnical Construction*, (9).

Taylor, E. H. (1991) The Khasab Self Spillway Embankment Dam. 17° ICOLD, Q.67, R.12, Vienne.

Taylor, G. (1968) *The Performance of Labyrinth Weirs*. PhD Thesis, University of Nottingham, Nottingham, England.

Todd, R. (1999) Spillway tainter gate failure at Folsom dam, California. *Waterpower Conference, ASCE, July 6–9, Las Vegas.*

Tozzi, M. J. (1992) *Characterization/Behavior of the Flow in Stepped Spillways*. Thesis of Doctorate, University of São Paulo, Brazil.

Tullis, J. P., et al. (1995) Design of labyrinth spillways. *PAP-782, Water Resources Research Laboratory, JHE*, 112(3), March.

USACE, United States Army Corps of Engineers (1965) HDS – Hydraulic Design of Spillways – Engineer Manual EM 1110–2–1603, Headquarters, Dept of the Army, Office of the Chief of Engineers. Available on the internet version .pdf.

USBR (1948) Studies of Crests for Overfall Dams, Boulder Canyon Project Final Reports, pt. VI, Hydraulic Investigations, Bulletin n. 3.

USBR (1952) *Discharge Coefficients for Irregular Overfall Spillways*. Engineering Monograph 9, Washington, DC.

USBR (1963) *Hydraulic Design of Reservoir Outlet Structures*. Engineering Monograph 1110–2–1602, Washington, DC.

USBR (1958/1960/1973/1974) *Design of Small Dams*. A Water Resources Technical Publication, Washington, DC, USA.

USBR (1980a) *Air-Water Flow in Hydraulic Structures.* Engineering Monograph 41, Washington, DC.

USBR (1980b) *Hydraulic Laboratory Techniques.*

USBR (1983) Dams and Public Safety. A Water Resources Technical Publication by Jansen, R. B. US Department of the Interior, Bureau of Reclamation.

USBR (1990) *Cavitation in Chutes and Spillways.* Engineering Monograph 42, Washington, DC.

USBR (1992) Alternatives for Enhancing Spillway Capacity Currently Being Pursued by the USBR. Vermeyen, T. and Mares, D.

USBR (2002) Hydraulic Design of Stepped Spillways. Final Report. Research Project 99FC800156. Denver, Colorado.

USBR (2015) Guidelines for Hydraulic Design of Stepped Spillways. Hydraulic Laboratory Report HL-2015-06.

USGS/Bodhaine, G. L. (1968) *Measurement of Peak Discharge at Culverts by Indirect Methods: Techniques of Water-Resources Investigations of the United States Geological Survey.* Book 3, Applications of Hydraulics, USA.

Vasconcelos, V., et al. (1986) The hydraulic performance of the surface spillway of paulo afonso IV plant. *XII Latin American Congress of Hydraulics, ABRH, São Paulo, Brazil.*

Veronese, A. (1937) Erosoni di fondo a valle di uno scarico. *Annali dei Lavori Publici*, 75, No. 9. Roma, Italy.

Vinogg, L. & Elstad, I. (2003) *Mechanical Equipment. Hydropower Development*, Vol. 12. Norwegian University of Science and Technology, Department of Hydraulic and Environmental Engineering.

Vischer, D. L. & Hager, W. H. (1998) *Dam Hydraulics.* ETH Zentrum, Zürich, Switzerland. John Wiley & Sons. Series in Water Resources Engineering.

Vischer, D. L. & Trucco, G. (1985) The Remodelling of the Spillway of Palagnedra.15° ICOLD, Q.59, R.8, Lausanne.

Wagner, W. E. (1956) Morning glory shaft spillways: Determination of pressure-controlled profiles. *Trans. ASCE*, 121, 345.

Wengler, R. P. (1982) The layout of mossyrock arch dam in Narrow Canyon. *International Symposium on General Arrangement of Dams in Narrow Gorges, CBDB, Rio de Janeiro, Brazil.*

Wenselowski, K. (2016) 2009 sayano-sushenskaya hydro disaster. *Annual Engineering Insurance Conference,* Toronto.

Whittaker, J. G. & Schleiss, A. (1984) *Scour Related to Energy Dissipators for High Head Structures*, Swiss Federal Institute of Technology – Zurich.

World Water (1979) Cavitation Casts Doubt on Karun Spillway Design. (old Reza Shah Kabir, actual Shahid Abbaspour). Comment p. 3. News p. 6., June.

Xavier, L. V. (2003) Non-conventional solution for the Itapebi hydroelectric power plant spillway. *25th SNGB, Salvador.*

Yuditskii, G. A. (1963) Hydrodynamic action of a discharge nappe on fragments of a rocky bed and conditions of breakage of this. *Izvestiya VNII Gidrotekhniki, Moskva, Leningrad.* LNEC, Translation n. 442, Pinto de Campos, J. A. Lisbon (1983).

Zanon, A. (1988) The karakaya hydro-electric plant on Euphrates River in Turkey. *Idrotecnica* (2), March–April.

Ziparro, V. J. & Hasen, H. (1993) *Davis's Handbook of Applied Hydraulics.* McGraw-Hill, Inc., New York, NY.

Appendix I
Spillways deterioration and rehabilitation

Throughout the text were presented famous cases of accidents. It should be noted that the "Deterioration of Spillways and Outlet Works" issue was addressed in **Q. 71** of the 18th ICOLD, Durban, **1994**, reported by Nelson Pinto, Consulting Engineer, Brazil. It is recommended that interested parties consult the report, a treatise on the subject, whose content covers:

Deterioration processes

- Foundation deterioration
- Concrete deterioration
- Hydraulic processes (cavitation, abrasion, erosion, sediment and floating debris, uplift)
- Equipment deterioration
- Upgrading of design criteria

Design to reduce the rate of deterioration

- Foundation
- Concrete deterioration
- Hydraulic processes (cavitation, abrasion, erosion, sediment and floating debris, uplift)
- Equipment
- Design to reestablish safety (raising the dam crest, lowering the spillway sill, lengthening the spillway crest, spillways over embankment dams, fuse plugs)

Operation and maintenance

Primary concerns – summary

Since that event, the subject has been treated, directly or indirectly, by ICOLD in the next congresses, as related here:

- **Q. 75** – Incidence and failures of dams, Florence (Italy), **1997**;
- **Q. 76** – The use of risk analysis to support dam safety decisions and management,
- **Q. 78** – Monitoring of dams and their foundation, and
- **Q. 79** – Gated spillways and other controlled release facilities and dam safety, Beijing (Chine), **2000**;
- **Q. 82** – Aging and rehabilitation of concrete and masonry dams and appurtenant works, Montreal (Canada) **2003**;

Q. 85 – Management of the downstream impacts of dam operation, and
Q. 86 – Safety of earth and rockfill dams, Barcelona (Spain), **2006**;
Q. 91 – Dam safety management, Brasília (Brazil), **2009**;
Q. 93 – Safety,
Q. 94 – Flood discharge and
Q. 95 – Aging and upgrading, Kyoto (Japan), **2012**;
Q. 97 – Spillway and
Q. 99 – Upgrading and reengineering of existing dams, Stavanger (Norway), **2015**;
Q. 101 – Safety and risk analysis,
Q. 102 – Geology and dams, and
Q. 103 – Small dams and levees, Vienna (Austria), **2018**.

This appendix presents only a summary of the theme rehabilitation of spillways to increase flow capacity, also a summary of erosion repairs as well as the equipment repairs. It is noteworthy that the accidents called attention to "forgotten" questions such as:

- the need to reevaluate the flooding of old dam spillways using more modern methods (PMP/PMF) and the larger dataset available;
- the need to realize periodic maintenance of gates and other mechanical equipment.

Remembering:

- in section 1.3, the spillway of the Mascarenhas de Moraes HPP (Furnas) was presented; this spillway had its design flood increased by 31%, from 10.400 m^3/s to 13.656 m^3/s, after 40 years of operation;
- in section 5.3 was presented the case of erosion in the spillway basin of the Coaracy Nunes HPP; Eletronorte reviewed the hydrological studies in 2014, and the flood of design was reduced by 40%, from 12.000 m^3/s to 7.213 m^3/s, after 38 years of operation.

Australia

Eildon dam, located on the Goulburn river, 140 km from Melbourne. Built between 1915 and 1929, it was modified in 1935 to increase its capacity. The "Great Eildon" was built between 1950 and 1955.

Figure AI.I Picture of Eildon Dam

The spillway is 33 m high and 60 m wide with a 435 m chute, with a ski jump at the end; three wagon gates, 20m wide by 9,3 m high, designed for 3.400 m³/s with a head of 9,0 m at the takeoff lip. Since 1955, the maximum discharge spilled was 500 m³/s.

The flood studies (PMF) were revised in 2003, obtaining a flood of 6.900 m³/s, 100% greater.

The owner, after conducting studies in a reduced model, decided to maintain the same spillway and raise the dam, once, to pass the new flood; the NA of the reservoir will be larger.

To pass the new flood, the project head was estimated at 14,10m (Hampton *et al.*, 2004). The spillway underwent structural modifications to meet the most intense turbulence conditions for the new project flow.

South Korea

Imha dam, on the Banbyeoncheon river, a tributary of the Nakdong river, 14 km from Andong, Gyeongsangbuk Province, in the East Region of South Korea. It was implemented between 1987 and 1991 with the following objectives: flood control, supply and power

Figure AI.2 Pictures of Imha Dam

generation (50 MW). It is a rockfill dam with earth core and is 515 m long and 73 m high. The spillway, on the left abutment, with four gates was designed for the flood of 4.800 m³/s, corresponding to 1,2 times the flood of TR = 200 years (3.800 m³/s). The maximum capacity was 5.350 m³/s (Lee *et al.*, 2008).

During the passage of the Rusa and Maemi hurricanes in 2002 and 2003, the floods reached 7,113 m³/s and 6,665 m³/s, respectively. The flood studies were then reviewed, reaching a value of 14,800 m³/s (PMP/PMF), almost three times higher.

The owner (K-Water), after conducting feasibility studies, decided to built another spillway on the right abutment, in a tunnel under the existing spillway, with an additional flow capacity of 8,000 m3/s. Three tunnels with a diameter of 15 m were adopted, with lengths of L1 = 379 m; L2 = 421 m; L3 = 462 m. According to the spillway tunnels were implanted during the years of 2008 and 2009. It is recorded that, according to these authors, the geology of the site is composed of a granite in a zone of fractured failure.

The second picture shows the control structure of the tunnel as well as a view of the alignment of the tunnels. The six gates have a width of 11,8 m and a height of 14,65 m (Lee *et al.*, 2008).

India

Hirakud dam (1957), on the Mahanadi river in the eastern part of the country, 15 km from Sambalpur, Orissa. Multipurpose reservoir, 55 km long, including flood control. Installed power 347,5 MW. The dam is 61m high and has a length of 4,8 km, plus 21 km of dikes. The spillway has a capacity of 42.450 m³/s, with 98 sluice gates – 64 bottom gates and 34 segment gates. An additional spillway for an additional 21,000 m³/s is approved.

Narayanapur dam (Srinivas, 2015): Rupture of the gate n. 5 on October 6, 2005.

Dam on the Krishna river in Siddapur, Bijapur District, Karnataka.

Figure A1.3 Hirakud Dam Localization Map

Figure A1.4 Weld rupture at beam joint with trunnion.

Height of the dam = 25,8m; L = 10,64 km; Q = 37.950 m³/s; 30 radial gates of 15 m × 12 m.

Rupture of the gate N. 5 on 10/6/2005. The works of fabrication and assembly of the new gate, as well as reinforcement of the others, were completed in April/2007.

Stoplog span 5.

Figure A1.4 (Continued)

Mexico

Considering the rupture of one of the gates of La Villita on May 31, 2013, CFE proceeded and concluded in 2014 a survey of its dams considered of greater risk (file available on the Internet). Some of the more significant cases from this report are presented here.

According to the CFE report, the head on the gate was 10,40 m. A discharge at break time of 740 m^3/s was estimated. The works of repair of the span 1 and reinforcement of the others ended in September 2014.

Figure A1.5 La Villita Spillway

Malpasso dam (Nezahualcóyotl), Grijalva river, Chiapas

The discharges spilled at the Malpaso dam between the years of 1967 and 1970 caused erosion of the slabs of the emergency spillway dissipation basin.

Figure A1.6 Malpaso Dam

Between the years of 2014 and 2015, the damages were repaired. The concrete chute was redone. All the gates were reinforced. The flow capacity was increased to 16.400 m³/s.

Figure AI.7 Malpaso Spillway

Peñitas dam

In the Peñitas dam, between 2015 and 2016, work was done on restoration of civil works and structural reinforcement of gates.

Figure AI.8 Peñitas Spillway

Infiernillo dam (1964)

The three spillway tunnels were designed for the 10.400 m³/s. Due to the events of 2013, hurricanes Ingrid and Manoel, the CFE reviewed flood studies in 2015, with a flood of 13.000 m³/s for a TR = 10.000 years. CFE designed and built an additional tunnel. In addition, CFE has developed aerator designs to combat cavitation problems similar to those previously reported in Hoover, Glenn Canyon and **Serre-Ponçon**.

Figure AI.9 Infernillo Dam

Appendix 2
Overflow dams

A brief record should be made of the overflow dams:

A   Rockfill overflow dams.
B   Overflow dams with the downstream face lined.
C   Overflow dams with stepped spillway.

This is an economical solution for certain cases of dams with small flows to be discharged. The subject has received attention from many researchers, and one can find in the bibliography some examples and sufficient material for another book.

Rockfill overflow dams

Examples of the first case are the Khasab dams on the Musandam Peninsula of Oman, presented in Taylor (1991), and Bastelos dam in Portugal for water supply, presented by Pinto and Martins (2011), "A self-spillway rockfill dam."

Bastelos dam, Figure A2.1, was completed in 1993 in Bastelos river, basin of the Douro river, Mogadouro Municipality.

This is a rockfill dam with bituminous concrete face, 22 m high (Figure A2.2). The material (2) constitutes the spillway of the dam, the detail of which can be seen in Figure A2.3. The design flood is of the order of 130 m³/s for a return period of 1.000 years.

It is assumed that the central zone of the dam, material 2 (Figures A2.2 and A2.3), functions as a free fall zone of the flow through the rockfill, which, for this purpose, has a relatively uniform granulometric distribution, ranging from 0,70 to 1,25 m. The outlet zone of the downstream spillway (material 1), 7,0 meters high, consists of relatively uniform rockfill with a diameter ranging from 1,25 to 2,50 m.

Hydraulic design and stability analysis are presented by Pinto and Martins (2011). Details on the hydraulic aspects of design and subject can be found in Martins (1981, 1986). The discharge capacity is given by $Q = cL\sqrt{2g}H_d^{3/2}$. Considering $c = 0.4$, a sill with a width L = 47,5 m was estimated to pass the design flood 130 m³/s under a head $H_d = 1,4$ m.

The following illustration (Figure A2.8) shows the location of the dams in the Northern Region of Portugal, including the Bastelos dam near Picote and Bemposta dams on the Douro river on the border with Spain.

Figure A2.1 Bastelos dam (Portugal). Location map and view of the dam and reservoir.

Overflow dams with the downstream face lined

The second example is the case of an embankment spillway with a very smooth downstream slope revetted by concrete slabs, extracted from Grishin (1982) – Figure A2.9. Grishin does not cite the name of the project but advises that the subject has been studied in detail by Gordienko.

According to Grishin, this project was subjected to a field test for $Q = 60$ m^3/s under a flow velocity of 23 m/s.

Users are advised to refer to references FEMA (2014), Chanson (2009), USBR (1992) and Pravdivets (1989), among others, available on the internet.

Overflow dams with stepped spillway

For the third case, there are examples of spillway dams using the practical solution in gabions. This is the usual solution for small dam spillway projects.

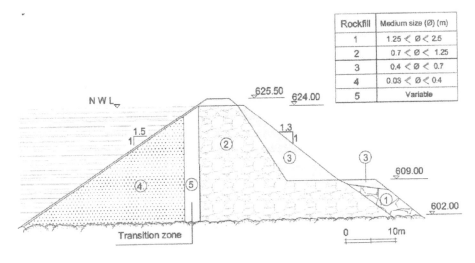

Rockfill	Medium size (∅) (m)
1	1.25 < ∅ < 2.5
2	0.7 < ∅ < 1.25
3	0.4 < ∅ < 0.7
4	0.03 < ∅ < 0.4
5	Variable

Figure A2.2 Bastelos dam. Typical section.

Figure A2.3 Bastelos dam. Rockfill spillway.

Figure A2.4 Bastelos dam. Intake.

Figure A2.5 Bastelos dam. Upstream slope.

Figure A2.6 Bastelos dam. Downstream view.

Figure A2.7 Bastelos dam. Downstream face.

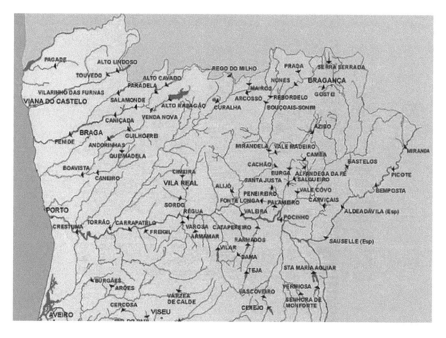

Figure A2.8 Map of the location of the dams in the Northern Region of Portugal.

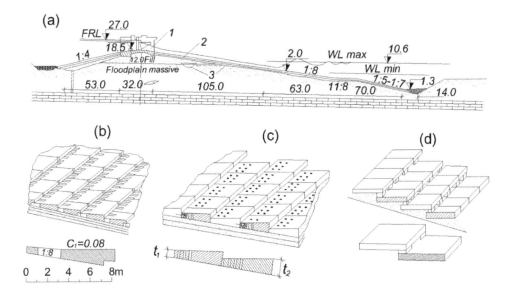

Figure A2.9 Embankent spillway. (1) Spillway. (2) Fill:(a) longitudinal section; (b) wedge-shaped
slabs with bondless cantilever; (c) same with flexible bonds, no cantilever; (d) flat
slabs of rectangular cross-section.

Source: (Grishin, 1982 – Figure 20.25).

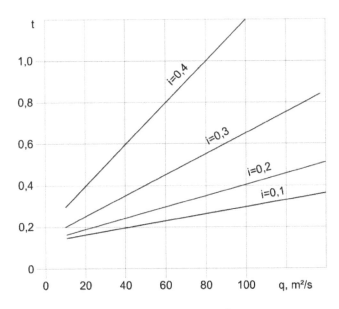

Figure A2.10 Lining thickness (t) × specific discharge (m²/s) × slope inclination (i).

Source: (Grishin, 1982 – Figure 20.25).

Figure A2.11 Gilboa dam spillway; 403 m long. One of the 19 reservoirs that supply water to NYC.

Source: En.wikepedia.org

Figure A2.12 New Croton spillway (NYC).

Source: Videoblocks.com

This solution in gabions is also used in projects of channels, protection of river banks and drainage in general, among other applications.

It is recommended that users consult, among others, the following publications:

- Chanson (93, 94, 95 and 2001);
- FAO: Food and Agriculture Organization of the United Nations. Small Dams and Weirs in Earth and Gabions Materials. AGL/MISC/32/2001;
- Gabion Energy Dissipators, Stephenson (1979);
- Maccaferri's catalogs.

Figure A2.13 Chinchilla spillway. $Q = 850$ m³/s. Width $= 410$ m. Height of the dam $= 14$ m.
Source: (Courtesy of Professor Hubert Chanson).

Figure A2.14 Gabion spillway (Maccaferri).

Index